Military English

전규호 · 정기주 공저

· 군대 실무분야로부터 군사 브리핑까지 단계적으로 구성
· 팝송을 포함하여 영어를 보다 쉽고 재미있게 학습하도록 구성
· 우리말 번역이 포함되어 보다 쉽게 학습 가능

도서출판 진영사

Military English

전규호 · 정기주 공저

2023년 9월 5일 초판 인쇄
2023년 9월 7일 초판 발행

발행인: 박진영
발행처: 도서출판 진영사
인천광역시 부평구 주부토로 236 인천테크노밸리 U1 지식산업센터 B동 1507호
전화: 032)505-4207
팩스: 032)505-4206
E-mail: 0183734207@hanmail.net
신고번호: 제2007-000001호(2007.01.19.)

ISBN 978-89-6541-622-7 93390
값: 17,000원

Military English

전규호 · 정기주 공저

도서출판 진영사

머리말

오늘날 우리 군은 지난 70여 년과 다름없이 당면한 위협으로부터 국가를 보위하는 사명을 지속 수행하는 가운데 기존의 전통적 위협에 비전통적 위협이 더해져 국가 간 협력의 중요성을 더욱 절감하고 있다. 특히, 대한민국은 지구상의 마지막 분단국으로서, 미국과의 혈맹관계를 70년간 지속 유지해 오고 있는 가운데 작전과 훈련을 함께하거나 단독 훈련시에도 연합작전을 가상하여 훈련하여 한국군에게 군사영어의 필요성을 더욱 증대시키고 있다.

군사영어는 일반적인 영어의 바탕 위에 군 고유의 활동과 관련된 어휘 및 전문용어, 브리핑, 작전명령 등과 같이 특화되고 전문화된 의사소통 방식을 취하고 있어, 이를 효과적으로 활용해야 의사소통이 자유로워진다.

이 책은 2개 PART 22개의 장으로 구성되어 있다. PART Ⅰ 실무 군사영어에서는 부대소개, 장병기본훈련, 연습과 훈련, 작전수행과정, 전방안내, 회의와 의전, 참모업무, 부대관리, 전투발전, 전사, 팝송을 구성하였다. 전투준비태세, 교육훈련, 전투발전, 부대관리, 전사 분야 대화에서 일반적으로 다루어지는 실무활동에 필요한 것들을 소개하였다. PART Ⅱ 군사브리핑에서는 자기소개, 북한의 위협과 군사력, 보병사단, 함대사령부, 비행단, 판문점, 도라전망대 지형설명, 인터뷰 예문, 일반적 표현, 회의진행, 유엔소개를 제시하였다. 외국군, 방문자들에게 핵심적인 사항을 간단, 명료하게 설명하는 데 도움을 주고자 하였다. 더불어 한국전쟁 73주년을 맞이하여 당시 유엔의 창설배경과 재한유엔기념공원 내용을 담아, 참전국의 감사함을 기억하고 잊지 않겠다는 의지를 담아 유엔소개를 포함하였다.

이 책은 연합 · 합동부대, 각 대학 · 사관학교에서 군사영어 교육 발전 및 교재개발에 기여하신 모든 분들의 노력이 축적된 결과물로서 모든 분들에게 감사의 마음을 전한다. 마지막으로 이 책의 출판을 맡아준 진영사 박진영 대표님께 감사드린다.

이 책이 군사영어를 공부하는 군사학도는 물론 한미연합분야 실무자, 군사영어에 관심 있는 사람들에게 도움이 되기를 바란다.

2023년 8월

집필자를 대표하여

전 규 호

Contents

PART Ⅰ. Practical Military English

Unit 1. Orientation of the Unit
Unit 2. Military Personnel Basic training
Unit 3. Exercise and Training
Unit 4. Operations Process
Unit 5. Front Area Guidance
Unit 6. Conference and Pretocol
Unit 7. Staff Responsibilities and Duties
Unit 8. Unit Management
Unit 9. Combat Development
Unit 10. War History
Unit 11. Learn English with Pop Songs

Unit 1. Orientation of the Unit

Maj Lee briefs Maj William on the general status.

B: Good morning. I'm Maj Lee, G-3 Operations Officer. First, I'd like to welcome you to the Division.
A: Glad to meet you. I'm Maj William.
B: This is the Division's Headquarters.
The staff offices are on the 1st floor, the Division Commander's office is on the 2nd floor, and the Command and Control Room is in the basement.
A: Could you show me around?
B: Sure, this way please.

== Related Expressions ==

- MG Hal La-San is the 52th Commander of the Division
- CofS G-1(Personnel), G-2(Intelligence), G-3(Operations), G-4(Logistics)
- All Sources Intelligence Center (ASIC)
- Chief of Staff, the 99th Division
- assistant division commander

Unit 1. 부대 소개

이 소령은 윌리엄 소령에게 부대 일반 현황을 소개한다.

B : 안녕하십니까? 저는 작전처 작전장교 이 소령입니다. 먼저 사단을 방문해 주신 것을 환영합니다.
A : 뵙게 되어 반갑습니다. 저는 윌리엄 소령입니다.
B : 이 건물이 우리 사단의 본청입니다. 참모부 사무실은 1층에, 사단장실은 2층에, 지휘통제실은 지하에 있습니다.
A : 안내해 주시겠습니까?
B : 물론입니다. 이쪽으로 가시죠.

== **관련 표현** ==

- 한라산 장군은 제52대 사단장이다.
- 인사참모, 정보참모, 작전참모, 군수참모
- ASIC(전 출처 정보 종합 센터)
- 99사단 참모장
- 작전 부사단장

1 History

B : The 99th Infantry Division was activated in 1948.
A : It has a long history.
B : Yes, that's right. We are proud of its long tradition in securing and maintaining freedom for our nation.
A : Could you tell me about a couple of well-known battles your division participated in?
B : Sure, our Division took part in the Hill Baekma battle and a lot of battle in the Vietnam War.
A : Oh, I see. By the way, what does your Division patch symbolize?
B : It symbolizes a huge whirlwind which represents boldness and fighting spirit.

== Related Expressions ==

- The unit moved to the present site in 1989.
- attachment, attached units
- to activate / to deactivate
- to achieve a brilliant victory in the battle at Hill Backma
- Presidential Unit Citation

1 연 혁

B : 제99 보병사단은 1948년에 창설되었습니다.
A : 부대의 역사가 오래되었군요.
B : 우리 부대가 한국전쟁 이전에 창설되었으며 자유를 쟁취하고 유지하는 데 오랜 전통을 지니고 있는데 대하여 자부심이 큽니다.
A : 사단이 참전한 유명한 전투는 무엇입니까?
B : 예, 우리 사단은 백마고지 전투와 월남전의 여러 전투에 참가했습니다.
A : 예, 그렇군요. 그런데 부대 표지는 무엇을 상징합니까?
B : 그것은 거대한 회오리를 상징하고, 과감성과 전투 정신을 의미합니다.

== 관련 표현 ==

- 부대는 1989년 현 위치로 이동했다.
- 배속, 배속부대
- 창설하다/해체하다
- 백마고지 전투에서 혁혁한 전공을 세우다
- 대통령 부대표창

2 Major Facilities

A : I appreciate your briefing on the Division.

B : My pleasure. Next, why don't we look around the facilities? The entrance over there with a guard post is the main gate. The grass field on the left is a parade ground. The building on the hill is the enlisted men's barrack.

A : What's the building on the right of the parade ground?

B : That is an amenities center. It has a mess hall, shower facilities, barbershops, and day rooms.

A : Troops ynder direct control of the Division are located together at Division HQ, aren't they?

B : Yes, they are. This is the Division's transportation unit. The buildings on the left side of the Division sign are the maintenance unit and the signal battalion.

== Related Expressions ==

- ammunition storage
- The Computer Center is a restricted area
- helipad
- supply storage
- POL(petroleum, oil and lubricants) storage

2 주요 시설

A : 사단 소개에 감사드립니다.
B : 천만에요. 다음은 부대 시설을 돌아보도록 하겠습니다. 저기에 보이는 초소가 사단 정문이고 왼쪽의 넓은 잔디밭이 연병장입니다. 그 위로 보이는 건물은 병사들의 내무반입니다.
A : 연병장 오른쪽에 보이는 건물은 무엇입니까?
B : 그것은 식당, 목욕탕, 이발소 그리고 휴게실 등이 있는 복지시설입니다.
A : 사단 직할부대는 사단 사령부 내에 같이 주둔하고 있죠?
B : 그렇습니다. 이곳이 사단 수송대입니다. 좌측의 사단 표지판이 있는 건물들이 정비대, 통신대대입니다.

== **관련 표현** ==

- 탄약고
- 전산실은 제한구역이다.
- 헬기장
- 보급품 창고
- 유류고

3 Missions and Organizations[1)]

A : What are the missions of the Division?

B : First, it is to secure the GOP and to defend the area of responsibility. Second, counter-infiltration operations and training.

A : Could you explain the organization of the Division?

B : The Division consists of six regiments : four infantry regiments and two field artillery regiment. Troops under the direct control of the Division are tank, engineer, signal, and reconnaissance battalions. Combat service support units are supply, transportation, maintenance, medical unit, and military police troops.

A : How many personnel and how much equipment are in the Division?

B : There are over 20,000 personnel in the Division. The major equipment includes 40 tanks, 210 pieces of artillery, and 400 vehicles.

== Related Expressions ==

- Division Commander's commanding emphasis
- homeland defense division
- mobilized division
- organic unit / task organization
- assignment / attachment

1) 김성우 · 정기주 『Low & High Level Military English』. pp.144~145.

3 임무 및 편성

A : 사단의 임무는 무엇입니까?
B : 첫째, GOP 경계 및 책임지역 방어입니다. 둘째, 대침투작전과 교육훈련입니다.
A : 사단의 편성에 대하여 설명해 주시겠습니까?
B : 사단은 4개 보병연대와 2개 포병연대로 편성되어 있습니다. 사단 직할대대로서 전차대대, 공병대대, 통신대대, 수색대대가 있습니다. 전투근무지원부대는 보급, 수송대, 정비대, 의무대, 군사경찰대가 있습니다.
A : 인원과 장비는 어떻습니까?
B : 인원은 20,000여 명입니다. 주요장비는 전차 40대, 화포 210문, 차량 400대입니다.

== 관련 표현 ==

- 사단장 지휘중점
- 향토 사단
- 동원 사단
- 편제부대 / 전투편성
- 예속 / 배속

4 Area of Operations[2)]

B : Maj William, we are located here at this point on the map.
A : It's not that far from Seoul, is it?
B : No, it is only 41km north of Seoul. Now, I'd like to explain the Division's area of operations. The width of our front line is 24km and the depth is 30km. The main road, Road No. 3 is well developed to the south and the north. The key terrains are Hill 475, Hill 380, and Munsan City.
A : There a lot of open terrains in this area.
B : It is easy for the enemy's mechanized forces to advance along Axis No. 3.

== Related Expressions ==

- natural obstacle
- defile depth
- front line
- depth
- area of operations(AO)
- area of interest(AOI)

2) 육군정보학교 『Military English』. pp.158~159.

4 작전지역

B : 윌리엄 소령, 우리가 있는 곳은 지도상 이 지점입니다.
A : 서울에서 그다지 멀지 않군요, 그렇죠?
B : 그렇습니다. 겨우 서울 북방 41km 정도 지점입니다. 그럼 사단의 작전지역을 설명하겠습니다. 사단은 정면 24km 종심 30km를 담당하고 있습니다. 3번 도로는 남북으로 잘 발달되어 있습니다. 중요 지형지물은 475고지, 380고지, 문산시 등입니다.
A : 개활한 지형이 많군요.
B : 3번 축선을 연하여 적 기계화 부대의 기동이 용이합니다.

== 관련 표현 ==

- 자연장애물
- 애로지역
- 정면
- 종심
- 작전지역
- 관심지역

5 Major Activities

A : I think your division is accomplishing a variety of missions.
B : Yes. Now 158th Regiment is conducting soldiers' basic training. 159th Regiment performs a security mission of the GOP along the front line.
A : We were tole that 157th Regiment is busy preparing for special training, right?
B : Yes, 157th Regiment is on RCT training in the field. The Artillery Regiment is conducting support activities for the training.
A : In addition to the various activities, the Division staff officers are planning to discuss Vision 2050.
B : We will discuss how to set the Division's objectives which can contribute to accomplishing Vision 2050.

== Related Expressions ==

- RCT(Regimental Combat team)
- to participate in an operations plan review conference at the Division HQ
- to check the management of barrier materials
- to repair and maintain tactical roads

5 주요활동

A : 사단은 다양한 임무들을 수행하고 있군요.
B : 예. 실제로 제158연대는 장병 기본훈련을 실시하고 있고 제159연대는 전방 GOP 경계 임무를 수행하고 있습니다.
A : 제157연대는 별도의 훈련을 준비하고 있다고 하는데, 맞습니까?
B : 예, 제157연대는 야외에서 연대 전투단 훈련을 실시하고 있습니다. 포병연대가 훈련을 지원하고 있습니다.
A : 다양한 부대 활동에 부가하여 사단 참모 장교들은 비전 2050에 대한 토의를 계획하고 있군요.
B : 우리는 비전 2050의 실현에 기여하기 위하여 사단의 목표 설정에 대하여 토의할 것입니다.

== 관련 표현 ==

- 연대 전투단
- 사단 사령부 작계검토회의에 참석하다
- 장병 자재 관리 실태를 확인하다
- 전술도로를 보수 및 정비하다

Unit 2. Military Personnel Basic training

Military Personnel Basic training.

A : What is the military personnel basic training?
B : It is essential training for military personnel to fight, survive, and win in the battlefield.
A : How many tasks does it have?
B : It has 21 tasks.
A : Do enlisted men have to master all of the common tasks?
B : Yes, that is basically right. Once the training level of a task reaches the required standards, the task can be omitted. We usually conduct the total system training with a combat models because all of the common tasks are related to each other.

== Related Expressions ==

- military occupational specialty(MOS)
- training evaluation
- training aids
- training guidance

Unit 2. 장병 기본훈련

장병 기본훈련

A : 장병 기본훈련에 관하여 설명해 주시겠습니까?
B : 장병 기본훈련은 장병들이 전장에서 싸우고 생존하고 승리하는데 필요한 필수적인 훈련입니다.
A : 총 몇 개의 세부 과제로 구성되어 있습니까?
B : 21의 세부 과제가 있습니다.
A : 모든 과제를 다 숙달해야 합니까?
B : 기본적으로는 그렇고, 수준이 달성된 과제는 생략할 수 있습니다. 모든 과제는 상호 연관성이 있으므로 훈련 시 유사 과제를 묶어서 전투모형 훈련을 실시합니다.

== 관련 표현 ==

- 군사 주특기
- 훈련 평가
- 교육 보조 재료
- 훈련 지침

1 Marksmanship

B : The shooters who have already zeroed their rifles are eligible for the day time qualification.
A : I see. In what conditions will night time fires be conducted?
B : They will be conducted by the soldiers who have successfully completed the daytime fires. We fire under two types of night conditions such as illuminated and non-illuminated environment, reflecting the real battlefield situation.
A : What is the minimum level of qualification for infantry soldiers?
B : At least 75% in day time firing and 50% in night time firing.

== Related Expressions ==

- battle sight alignment(battle sight zero)
- non-illumination
- weapons cleaning equipment
- standards for Army marksmanship
- preliminary rifle instructions(PRI)
- target detection training

1 사격술 훈련

B : 주간 자격사격은 전투가능자 영점을 조정한 사수만이 실시 가능합니다.
A : 예, 그렇군요. 야간 사격은 어떤 조건 하에서 실시합니까?
B : 주간 사격에 합격한 사수가 실시하며 전장실상을 고려하여 무조명, 유조명 조건으로 구분하여 실시합니다.
A : 보병 전투병의 자격사격 합격수준은 어느 정도입니까?
B : 주간 사격은 75%, 야간 사격은 50%이상입니다.

== 관련 표현 ==

- 전투가능자 영점
- 무조명
- 병기손질 도구
- 육군사격 기준표
- 사격술 예비훈련
- 표적탐지훈련

2 Daylight Movement

A : What is the purpose of daylight movement training?
B : It is to enable the soldiers to approach the objective promptly under the enemy's observation and fire.
A : What kind of ways; are there in daylight movement?
B : There are two ways; crawl and rush. The crawl has three ways; low crawl, high crawl, and applied crawl.
A : When do you do the high crawl?
B : First, when we are provided with cover and concealment by the terrain. Second, when visibility is limited. And third, when speed is required.
A : What is the best way to move in open areas?
B : It is rush.

== Related Expressions ==

- cover and concealment
- secrecy, stealth
- furtive movement, stealthy movement
- live fire
- surprise attack from an unexpected direction

2 주간이동

A : 주간이동 훈련의 목적은 무엇입니까?
B : 적의 관측과 사격 상황하에서 신속하게 목표로 접근하는 것입니다.
A : 주간 이동의 방법은 어떠한 것들이 있습니까?
B : 포복과 약진 2가지로 분류됩니다. 포복에는 높은 포복, 낮은 포복, 응용포복 등 3가지가 있습니다.
A : 높은 포복은 어떤 경우에 실시합니까?
B : 첫째, 지형상 은폐 · 엄폐를 제공받을 수 있을 때 둘째, 시계가 제한될 때 셋째, 속도가 요구될 때 사용됩니다.
A : 개활지를 이동하는 최선의 방법은 무엇입니까?
B : 약진입니다.

== 관련 표현 ==

- 엄폐 / 은폐
- 기도비닉
- 은밀한 이동
- 실사격
- 예기치 않은 방향으로부터의 기습 공격

3 Information Collection, Report and Taking Actions

A : Could you tell me about the information collection training?

B : We divide the trainees into two opposing groups (friend and foe). Two members of the friendly force will conduct a security mission while the enemy tries to approach the guard post.

A : How do the guards report the information when they find the enemy?

B : They should report it including three factory : who, what, where and when

A : What kind of actions are required?

B : They should be able to canalize the enemy into point-blank range and neutralize them with rifles and hand grenades, but capturing them is the best option, if possible.

== Related Expressions ==

- load bearing equipment(LBE)
- immediate action
- binoculars
- opposing forces(OPFOR)
- blank ammunition
- treatment and management of prisoners of war(POW)

3 첩보수집, 보고 및 조치

A : 첩보수집 훈련은 어떻게 실시됩니까?
B : 쌍방(적군 · 아군)으로 조를 편성합니다. 아군 2명은 지정된 초소에서 경계임무를 수행하고 적군은 초소로 접근합니다.
A : 적군을 발견한 초병은 첩보 보고를 어떻게 합니까?
B : 적군을 발견한 초병은 "누가, 무엇을, 어디서, 언제" 사항을 포함하여 보고합니다.
A : 어떤 조치가 요구됩니까?
B : 지근거리까지 유인하여 사격, 수류탄 투척으로 제압하는 능력이 요구됩니다. 그러나 가능하다면 포획하는 것이 최상입니다.

== **관련 표현** ==

- 단독군장
- 즉각 조치
- 쌍안경
- 대항군
- 공포탄
- 포로 취급 및 관리

4 First Aid Treatment

A : What is the key factor for successful first aid treatment?

B : The key is to practice the four phases of first aid action to save one's life.

A : What are the four phases of first aid action?

B : It consists of four steps. First, clear the air passage, second stop bleeding. Third, prevent shock and continue treatment. Finally, protect wounds.

A : Could you tell me how to stop bleeding?

B : We have two methods, one is direct pressure, the other is using a constriction band.

A : Is there anything else to understand about this subject?

B : Yes, there are two more points. They are how to use a splint for fractures and how to move the patient.

== Related Expressions ==

- constriction band
- wounded in action(WIA)
- head and chest wounds

4 환자 응급 처치

A : 성공적인 환자 응급처치의 핵심은 무엇입니까?
B : 환자의 생명을 구하기 위하여 구명 4단계를 숙달하는 것입니다.
A : 응급처치 구명 4단계는 무엇입니까?
B : 응급처치 구명 4단계는, 1단계 : 기도 유지, 2단계 : 지혈, 3단계 : 쇼크방지 및 지속적인 치료, 4단계 : 상처 보호입니다.
A : 지혈방법에는 어떤 것이 있습니까?
B : 2가지 방법이 있는데 직접압박법, 지혈대 사용법이 있습니다.
A : 이 훈련과제에서 추가로 알아야 할 내용이 있습니까?
B : 예, 2가지 요소가 있습니다. 골절환자의 부목법, 환자운반법이 있습니다.

== **관련 표현** ==

- 압박붕대
- 부상자
- 두부 및 흉부손상

5 Mission Oriented Protective Posture (MOPP)

A : How many levels are there in the MOPP and how do you divide them?
B : There are five levels from zero to four. Each level specifies what equipment must be worn and how to act.
A : What is the most important thing you should do prior to a nuclear attack?
B : Digging-in is the most important.
A : What kind of action is needed in case of fallout contamination?
B : We should stay in the trench wearing a mask and a rain coat. It is very important that we are staying in a covered trench.

== Related Expressions ==

- Do not expose your skin to heat radiation.
- radioactive rays
- boot covers
- nuclear, biological, chemical (NBC) report

5 임무형보호태세

A : 임무형보호태세는 몇 단계가 있으며 어떻게 구분됩니까?
B : 0단계에서 4단계까지 5단계로 구분됩니다. 각 단계별로 착용 장비 및 행동이 명시됩니다.
A : 핵 공격전 행동 시 가장 중요한 것은 무엇입니까?
B : 호를 구축하는 것이 가장 중요합니다.
A : 방사능에 오염된 낙진 시에는 어떤 조치가 필요합니까?
B : 방독면과 우의를 착용하고 호에서 대기해야 합니다. 유개호에 대기하는 동안에 낙진에 신체가 노출되지 않도록 하는 것은 매우 중요합니다.

== 관련 표현 ==

- 복사열에 피부를 노출하지 말라.
- 방사선
- 전투화 덮개
- NBC(화생방) 보고

6 Hand Grenade Attack

A : What effects do we expect to achieve when attacking with hand grenades?

B : First, it can reinforce the effectiveness of rifle fire. Second, it forms a comparatively wider kill zone against the enemy. Third, it makes the enemy frightened while building morale among our forces.

A : What are the major targets in hand grenade training?

B : Exposed troops, a covered trench, foxhole or communication trench could all be targets.

== Related Expressions ==

- dummy hand-grenade
- smoke hand grenade
- effective casualty radius
- how to grip a hand grenade

6 수류탄 공격

A : 수류탄으로 적 공격 시 어떤 효과가 있습니까?
B : 첫째, 소총 사격 효과를 강화할 수 있습니다. 둘째, 비교적 넓은 지역의 적을 살상할 수 있습니다. 셋째, 적에게 공포감을 주고 반대로 아군의 사기를 높입니다.
A : 수류탄 공격 훈련시 주요 표적은 무엇입니까?
B : 노출된 병력, 엄폐호, 개인호 및 교통호를 선정합니다.

== 관련 표현 ==

- 연습용 수류탄
- 연막 수류탄
- 유효 살상 반경
- 수류탄 파지 요령

Unit 3. Exercise and Training

Maj Lee takes Maj William to the tactical training site.

A : Does the US Army also emphasize the military personnel basic training?
B : Of course, because the military personnel basic training is the basis of unit tactical training.
A : I'm going to escort you to the training site now.
B : I'm interested in the Korean soldiers' unique training model since we're in a combined forces system.
A : I see. So far I know you din't have any training for counter-infiltration operations which the ROK Army has.
B : I'm getting more interested. This is my first visit to a tactical training site of the ROK Army.
A : Good. Let me explain the training procedure as we observe the Ulchi Freedom Shield.

Unit 3. 연습 및 훈련

이 소령은 윌리엄 소령을 전술 훈련장으로 안내한다.

A : 미 육군에서도 장병 기본훈련은 중요시 되겠지요?
B : 물론입니다. 장병 기본훈련은 부대 전술훈련을 위한 기초가 되는 것이기 때문입니다.
A : 지금부터 부대 전술훈련 현장으로 안내하겠습니다.
B : 우리가 연합군 체제이기에 한국군 특유의 훈련 형태가 궁금하군요.
A : 그러시군요. 대침투작전 같은 것은 미군의 전술훈련에는 없는 것으로 알고 있습니다.
B : 한국 육군의 전술훈련을 참관하는 것이 처음이라서 더욱 궁금해 집니다.
A : 그럼 을지 자유의 방패 훈련부터 참관하면서 훈련 절차를 설명하겠습니다.

1 Ulchi Freedom Shield Exercise

A : What is the purpose of this exercise?
B : The purpose is to check the overall combat readiness and necessary measures including mobilization readiness.
A : Then, the exercise is not limited to the MND, is it?
B : No, it isn't Each department of the government takes part in the exercise.
A : I think future warfare would be a total war, and we need to prepare for that.
B : That's right. It is combat forces including standing troops that win the battles by direct engagement against the enemy. But effective mobilization, defense industry, and civil order in the rear area are as important as combat forces.

== Related Expressions ==

- integrated defense system with civil, government and military
- noncombatant evacuation operations(NEO)
- to check mobilization resources
- mobilization of personnel, materiel, and facility
- Ministry of National Defense(MND)

1 을지 자유의 방패 연습

A : 금번 훈련의 목적은 무엇입니까?

B : 이 훈련을 통하여 동원태세를 포함한 전체적인 전쟁 준비태세와 보완해야 할 사항을 점검하는 것입니다.

A : 그러면, 국방부에 국한된 훈련은 아니겠습니다.

B : 예. 정부 각 부서가 참가하게 됩니다.

A : 장차전은 총력전에 대한 대비가 될텐데요?

B : 그렇습니다. 직접 교전을 하여 승리를 달성하는 것은 상비군을 포함한 전투부대입니다. 그러나 효과적인 동원, 방위산업, 후방 치안 유지 또한 전투부대 못지 않게 중요합니다.

== 관련 표현 ==

- 민 · 관 · 군 통합 방위체제
- 비전투원 후송 작전
- 동원 자원을 확인하다
- 인원, 물자 및 시설 동원
- 대한민국 국방부

2 Regimental Combat Team Training

A : Can you tell me the organization of the regimental combat team?
B : Yes. It consists of an infantry regiment and some attached units such as armored units, artillery, and engineers.
A : What are the considerations for the organization of a regimental combat team?
B : They are METT-TC
A : What are the characteristics of the regimental combat team?
B : It can perform its missions independently and be commanded by the assistant division commander, if necessary

== Related Expressions ==

- METT-TC
  * Mission, Enemy, Terrain and Weather, Troops available, Time available, Civil considerations
- integrated employment of fire power
- tactical march
- to establish a communication
- to employ supporting and attached units effectively

2 연대 전투단 훈련

A : 연대 전투단은 어떻게 편성되어 있습니까?
B : 연대 전투단은 보병연대와 그에 배속된 기갑부대, 포병 및 공병 등입니다.
A : 연대 전투단 편성시 고려사항은 무엇입니까?
B : 그것은 METT-TC입니다.
A : 연대 전투단의 특징은 무엇입니까?
B : 연대 전투단은 독립적으로 임무를 수행하며 경우에 따라 부사단장에 의한 지휘도 가능합니다.

== 관련 표현 ==

- 임무변수(구, 전술적 고려요소)
  * 임무, 적, 지형 및 기상, 가용 부대, 가용 시간, 민간요소
- 통합화력 운용
- 전술적 행군
- 통신망을 구성하다
- 지원 및 배속부대를 효과적으로 운용하다.

3 Counter-infiltration Operations

A : What do you estimate the probable enemy's infiltration routes to be?
B : We estimate that they will use the three routes they used for infiltration before in the DMZ.
A : What could be the possible objectives of infiltrators?
B : The patriot battalion and the airfield in our division's rear area could be their important objectives.
A : What is your plan to detect the enemy infiltrators and to destroy them?
B : We have five interdiction lines and four blockade lines based on the Division's operations plan.
A : How about the task organization?
B : We have assigned five areas of responsibility to each regiment in depth, then each regiment conducts its own task organization in the area.

== Related Expressions ==

- training for entrapping the enemy within the effective range of fire
- to preserve the initial point
- to discover the enemy's trace
- to search and destroy
- a combined interrogation team

3 대침투 작전

A : 적의 예상 침투로는 어떻게 판단하고 있습니까?
B : 현 휴전선 지역의 기존 침투로 3곳을 이용할 것으로 판단됩니다.
A : 침투하는 적의 목표는 무엇입니까?
B : 패트리어트 대대와 사단의 후방지역에 위치한 비행장이 적의 중요한 목표가 되리라고 판단하고 있습니다.
A : 침투하는 적을 포착, 격멸하기 위해 수립된 계획은 무엇입니까?
B : 사단은 5개의 차단선과 4개의 봉쇄선을 계획하였습니다.
A : 전투편성은?
B : 각 보병연대를 종심으로 5개 분구로 책임지역을 부여하고, 각 연대는 지역별로 전투편성을 실시하였습니다.

== 관련 표현 ==

- 지근거리 유도 훈련
- 원점을 보존하다
- 적의 흔적을 발견하다
- 탐색 및 격멸하다
- 연합 심문조

4 CPM-X

A : What is the CPM-X?

B : It is the Command Post Movement Exercise for the commanders and staff officers among higher and lower headquarters while maintaining the communication network.

A : How is it conducted?

B : After being completely prepared for the movement exercise through the Rock Drill, we conduct the CPM-X.

A : How is it controlled?

B : Controllers and umpires thoroughly prepare for it by war game.

A : What is the most important thing in the exercise?

B : Establishing a firm communication network among the echelons is the most essential part of the exercise.

== Related Expressions ==

- to guard the CP
- main CP/alternate CP/tactical CP
- After Action Review(AAR)
- to organize opposing force

4 지휘조 기동훈련

A : 지휘조 기동훈련은 무슨 훈련입니까?

B : 지휘조기동훈련은 사령부(지휘소)내 및 상 · 하급 사령부(지휘소)간에 지휘관과참모들이 통신을 유지하면서 실시하는 야외 훈련입니다.

A : 지휘조 기동훈련은 어떻게 실시됩니까?

B : 지휘조 기동훈련은 부대 내에서 먼저 지휘소 연습을 통해서 충분히 준비한 후 야외 훈련을 실시합니다.

A : 지휘조 기동훈련의 통제는 어떻게 합니까?

B : 통제관과 심판관들은 워게임을 통하여 철저하게 통제를 준비합니다.

A : 지휘조 기동훈련에서 가장 중요한 사항은 무엇입니까?

B : 훈련 간 각 제대 간의 통신망 확립이 가장 중요한 사항입니다.

== 관련 표현 ==

- 지휘소를 경계하다
- 주 지휘소/예비 지휘소/전술 지휘소
- 사후검토, 강평
- 대항군을 편성하다

5 Battalion Tactical Training

A : What is the period of a training cycle?
B : A training cycle is a period from the start of the individual training to the end of battalion tactical training.
A : Then, how is the battalion tactical training conducted?
B : First, the training plan is made according to the higher commander's training intent. Officers conduct their training, teaching demonstrations, CPXs and Command Post Movement Exercises while NCOs conduct team, squad and platoon training. After that, we all together conduct the Battalion field maneuver exercise.
A : What's the purpose of the exercise?
B : It is to practice the primary tasks and combat skills which are essential in the future battle field to accomplish the unit mission.

== Related Expressions ==

- to defeat the enemy
- to draw lessons learned and apply them to the next training exercise
- tactical assembly area(TAA)
- OPLAN implementation exercise

5 대대 전술훈련

A : 훈련주기는 어떻게 됩니까?
B : 개인훈련으로부터 대대 전술훈련을 완성하는 것을 1개 주기로 합니다.
A : 그러면 대대 전술훈련은 어떻게 실시됩니까?
B : 우선 상급 지휘관의 훈련 복안서에 따라 훈련계획을 작성합니다. 그리고 장교는 교육 및 연구강의, 지휘소 훈련, 지휘조 기동훈련을 실시하며 부사관은 팀, 분대, 소대훈련을 실시합니다. 마지막으로 야외 기동훈련을 합니다.
A : 대대 전술훈련의 목적은 무엇입니까?
B : 장차전에서 대대가 수행해야 할 임무 완수에 요구되는 주요 수행 과업 및 전투기술을 숙달하는 것입니다.

== 관련 표현 ==

- 적을 격퇴하다
- 훈련 교훈을 도출하여 차기 훈련계획에 반영하다
- 전술 집결지
- 작전계획 시행 훈련

6 Close Air Support(CAS)

A : Are there any CAS sorties assigned to your division during the training?
B : Yes, ten sorties a day are assigned to our division.
A : Are there any requests from the subordinated units?
B : Sure, we have already used three sorties against a target of one enemy tank battalion which appeared in front of our 157th Regiment.
A : Do you have any other fire support assets?
B : Yes, we do. However, we estimate the tank battalion is a big threat to us, and the Air Force support will be more effective than Army aviation or tank attack So, we requested the CAS immediately.
A : The ground air control officer did a very good job, didn't he?
B : Yes, he sure did.

== Related Expressions ==

- air assault
- secure local air superiority
- forward air controller contrils air sorties
- attack helicopter
- air space control plan

6 근접 항공 지원

A : 금번 훈련간 근접 항공 지원자산이 할당되어 있습니까?
B : 예, 사단에는 기간동안 일일 10소티씩 할당되어 있습니다.
A : 예하대 요청은 있습니까?
B : 예, 157연대 전방지역에 출현한 적 전차 1개대대에 대하여 3소티를 이미 사용했습니다.
A : 다른 화력지원 자산은 없습니까?
B : 아니요, 있습니다. 그러나 이번 상황은 우리에게 커다란 위협요소로 판단되었고 육군항공이나 전차 공격부대 운용보다는 공군 지원이 효과적이라 판단하여 즉각 CAS를 요청한 것입니다.
A : 지상 항공 통제장교가 큰 역할을 하였겠지요?
B : 예, 그렇습니다.

== 관련 표현 ==

- 공중강습
- 국지적인 공중 우세권 확보
- 전방 항공 통제관이 항공기를 유도하다
- 공격 헬기
- 공역 통제 계획

Unit 4. Operations Process

Maj Lee introduces the Operations Process to Maj William.

A : What is the purpose of the Operations Process?
B : We follow the procedure to accomplish the assigned missions effectively.
A : Please explain the Operations Process.
B : It is conducted in four phases, continuously and systematically. Phase one is the planning. Phase two is the preparation for the operations. Phase three is the execution.
But there are some difficulties applying these phases to all levels of echelon.

== Related Expressions ==

- efficient management of time, personnel, equipment and resources
- dominant factors
- joint efforts by the commander and his staff
- flexible planning

Unit 4. 작전수행 과정

이 소령은 윌리엄 소령에게 작전수행 과정을 소개한다.

A : 작전수행 과정의 목적은 무엇입니까?
B : 부여된 임무를 효과적으로 달성하기 위하여 적용합니다.
A : 작전수행 과정 절차를 설명해 주십시오.
B : 3단계에 걸쳐 연속적, 체계적으로 진행됩니다. 제1단계는 계획수립입니다. 제2단계는 작전준비입니다. 제3단계는 작전실시입니다.
그러나 모든 제대에 이 절차를 적용하는데는 어려움이 있습니다.

== 관련 표현 ==

- 시간, 인원, 장비 및 자원의 효율적 관리
- 지배적인 요소
- 지휘관 및 참모의 공동 노력
- 융통성 있는 계획 수립

1 Planing

(1) Situation Evaluation

A : What activities are in the planing phase?
B : They are evaluation of the situation, outlining of the operations, development of COAs, and completion of the planing.
A : What is the most important factor in the first phase of the evaluation of a situation?
B : It is mission analysis.
A : Could you explain to me the procedure for mission analysis?
B : Yes. First, we are to identify the specified tasks in the given missions. Second, to establish the purpose of the operations. Third, to draw implied tasks.

1 계획수립

(1) 상황평가

A : 계획수립 단계에는 무슨 활동이 포함됩니까?
B : 이 단계에서는 상황평가, 작전구상, 방책발전, 계획완성 등이 포함됩니다.
A : 첫 단계인 상황평가 단계에서 가장 중요한 사항은 무엇입니까?
B : 임무분석이 가장 중요합니다.
A : 임무분석 절차에 대하여 설명해 주시겠습니까?
B : 예, 첫째, 상급부대에서 하달된 명시된 과업을 식별합니다. 둘째, 작전목적을 설정합니다. 셋째, 추정된 과업을 염출하게 됩니다.

(2) Outlining Operations

A : Who takes initiative in outlining the operation?
B : The division commander leads this phase, and then issues the planning guidance.
A : What does the planning guidance include?
B : First, the commander will present his initial intent for the operation to his staff officers. Second, he issues his guidance for tactical deception. Third, he issues his priority intelligence requirements(PIR). Fourth, he gives guidance for the time plan.
A : What is commander's initial intent?
B : It is to specify his operational purpose, the method of unit employment, and the end state.

== Related Expressions ==

- to achieve relative superiority in combat power at the decisive time and place
- events that require early decision
- to fix and contain
- essential elements of information(EEI)
- line of contant(LC) / line of departure(LD)
- to maintain the momentum of attack

(2) 작전구상

A : 작전구상은 누가 주도합니까?
B : 사단장이 주도하며 작전구상 후 계획 지침을 하달하게 됩니다.
A : 구체적으로 어떤 내용이 포함됩니까?
B : 첫째, 사단장은 최초 지휘관 의도를 참모들에게 제시합니다. 둘째, 전술적 기만 지침을 지시합니다. 셋째, 지휘관의 중요 첩보요구를 하달합니다. 넷째, 시간 사용 지침 하달입니다.
A : 최초 지휘관 의도란 무엇입니까?
B : 사단장의 작전목적, 부대운용 방법, 최종상태를 명시하는 것입니다.

== **관련 표현** ==

- 결정적인 시간과 장소에서 상대적 전투력 우세를 달성하다
- 조기 결심이 요구되는 사항
- 고착 및 견제하다
- 첩보 기본 요소
- 접촉선/공격 개시선
- 공격 기세를 유지하다

(3) COA Development

A : What activities should be conducted in the phase of COA development?
B : They are the analysis of AO, threat analysis, COA development, COAs analysis, comparison of COAs and COA selection.
A : Does the analysis of AO mean the IPB?
B : Yes. We analyze key terrains and avenues of approach through the IPB.
A : How do you develop the COAs?
B : First, we select and compare the enemy's probable COAs. Second, analyze the enemy's strengths and weaknesses. Third, select friendly force's COA.
A : How do you compare COAs?
B : In general, we compare them by analyzing advantages and disadvantages of COAs.

== Related Expressions ==

- course of action(COA)
- intelligence preparation of the battlefield(IPB)
- evaluation of the battlefield
- evaluation of enemy's ability
- area of operations(AO)
- selection of the best course of action
- favorable for employment of infiltration forces

(3) 방책발전

A : 방책발전에서 실시되는 활동에는 어떤 것이 있습니까?
B : 작전지역 분석, 위협 분석, 방책 수립, 방책 분석, 방책 비교 및 선정입니다.
A : 작전지역 분석이란 IPB를 말하는 것입니까?
B : 그렇습니다. IPB를 통하여 중요지형지물 및 접근로를 분석하게 됩니다.
A : 방책 수립은 어떻게 합니까?
B : 첫째, 적 방책을 선정 및 비교합니다. 둘째, 적의 강 · 약점을 분석합니다. 셋째, 아 방책을 선정하게 됩니다.
A : 방책 비교 방법에는 어떤 것이 있습니까?
B : 일반적으로 장 · 단점에 의한 비교를 실시합니다.

== **관련 표현** ==

- 방책
- 전장 정보 분석
- 전장 평가
- 적 능력 평가
- 작전지역
- 최선의 방책 선정
- 침투부대 운용에 유리

(4) Completion

A : What does the completed plan include?
B : It includes the operational plan and its annexes.
A : What kind of annexes do you have?
B : The annexes are task organizations, operation overlay, fire support, intelligence, engineer, signal, service support, and deception.
A : What does the operation overlay show?
B : It shows a scheme of maneuver, an integrated fire support, and a combat service support plan.

== Related Expressions ==

- operation overlay
- Annex B(Intelligence)
- field SOP(standing operating procedure)
- scheme of maneuver
- task organization
- mobile ammunition supply point(MASP)

(4) 계획 완성

A : 완성된 계획에는 무엇이 포함됩니까?
B : 완성된 계획에는 기본문과 부록이 포함됩니다..
A : 부록은 어떤 것들이 있습니까?
B : 부록에는 전투편성, 작전투명도, 화력지원 계획, 정보, 공병, 통신, 근무지원, 기만 등이 포함됩니다.
A : 작전투명도는 어떤 내용을 나타냅니까?
B : 기동계획, 통합 화력지원, 작전지속지원 계획 등을 묘사합니다.

== 관련 표현 ==

- 작전투명도
  * 작전계획 또는 작전명령을 군대부호와 전술적 통제수단을 사용하여 도식한 투명도
- 부록 B(정보)
- 야전예규
- 기동계획
- 전투편성
- 이동 탄약보급소

(5) Issuance of Order

A : Does issuing order mean the phase assigning missions to the subordinate units?
B : Yes. Based on the order, the subordinate units begin taking actions.
A : What are the important points when you issue orders?
B : We assign missions to subordinate units, make sure that they understand them, and allocate the time for coordination.
A : How do you issue orders to your subordinate units?
B : We normally issue them directly or by couriers to the subordinate commanders.
A : Could you explain the time provided for coordination?
B : The subordinate units are given sufficient time for close coordination.

== Related Expressions ==

- to avoid misunderstanding and confusion by using clear and concise plans and orders
- fragmentary order(FRAGO)
- battlefield surveillance measures
- direction of attack

(5) 명령 하달

A : 명령하달이란 예하부대에 임무를 부여하는 단계를 말합니까?
B : 그렇습니다. 이 명령에 근거하여 예하부대의 활동이 시작됩니다.
A : 명령하달에는 어떤 내용이 포함됩니까?
B : 임무부여, 임무확인, 협조시간 부여가 포함됩니다.
A : 명령하달 방법에는 어떤 형태가 있습니까?
B : 통상 예하 지휘관에게 직접 하달하거나 전령을 운용하기도 합니다.
A : 협조시간 부여란 무엇입니까?
B : 명령수령 후 예하부대가 상호 긴밀히 협조하는 시간을 부여합니다.

== 관련 표현 ==

- 간명한 계획과 명령으로 착오와 혼란을 방지하다
- 단편 명령
  * 이미 하달한 명령 내용의 적시적인 수정이 요수 될 때 변경사항을 지시하는 명령
- 전장 감시 수단
- 공격방향 화살표

2 Preparation of Operations

A : What are the major activities in the preparation of operations?
B : They are coordination, confirmations/supervision, and rehearsal in this phase.
A : How do you supervise the subordinate units?
B : The division commander can either visit the subordinate units or make the staff members supervise them.
A : And the rehearsals can be done when time permits, right?
B : That's not always the case. The rehearsals for the critical phases of the operations should be done whenever possible.

== Related Expressions ==

- mission briefing
- warning order
- fundamentals of offensive operations
- tactical control measures

2 작전준비

A : 작전준비간 실시되는 주요 활동은 무엇입니까?
B : 협조, 확인/감독, 예행연습 등입니다.
A : 예하 부대를 어떻게 감독합니까?
B : 사단장이 직접 예하 부대를 방문하여 실시하거나 참모조직을 활용하여 실시할 수도 있습니다.
A : 예행연습은 시간적 여유가 있을 때에만 실시하겠군요.
B : 반드시 그렇지는 않습니다. 작전의 중요 국면은 가능한 한 예행연습을 실시합니다.

== **관련 표현** ==

- 임무 브리핑
- 준비 명령
- 공격 작전 준칙
- 전술적 통제 수단

  * 모든 작전에 공통적으로 적용 할 수 있는 의사소통의 주요 도구로서 부대의 활동과 전투수행 기능을 제한, 조정, 협조, 통합시키기 위해 사용함

3 Execution

A : This would be the most important phase in the unit command procedure, right?
B : That's correct. This is a very important phase for the success of operations.
A : What activities are conducted in this phase?
B : It is an ongoing cycle of assessing situation, making decisions, and issuing orders.
A : What are the roles of the commander and staff?
B : The staff should make prompt and precise estimates of the situation. Based on the evaluation, they will recommend an appropriate course of action. The commander should make timely decisions.

== Related Expressions ==

- to force the enemy to disperse
- to commit the reserve
- contingency plan(CONPLAN)
- rear area operations

3 작전실시

A : 이것은 부대 지휘 절차의 가장 중요한 단계가 되겠군요.
B : 맞습니다. 이 단계는 작전의 성패를 결정합니다.
A : 이 단계에서는 어떤 활동이 수행됩니까?
B : 상황 판단하고, 결심하고, 명령을 하달하는 연속적인 과정입니다.
A : 지휘관과 참모의 역할은 무엇입니까?
B : 참모는 신속하고 정확한 상황판단을 해야 합니다. 참모는 상황평가에 근거하여 이루어져야 할 적절한 조치를 건의합니다. 지휘관은 적시적인 결심을 해야 합니다.

== **관련 표현** ==

- 적 부대의 분산을 강요하다
- 예비대를 투입하다
- 우발 계획
- 후방 지역 작전

Unit 5. Front Area Guidance

Maj William visits a unit in front area

A : I've learned a lot about the training for military personnel basic and leaders.

B : I hope it helps you better understand the ROK Army.

A : I really appreciate your king explanation.

B : Now, we're going to visit a division in front area.

A : What places of the division will we visit?

B : We are going to visit an OP, GOP, and then take a look at some barriers and minefields. We will also look at an invasion tunnel dug by North Korea.

Unit 5. 전방안내

윌리엄 소령은 전방부대를 방문한다.

A : 장병 교육훈련과 간부훈련에 관하여 많은 것을 배웠습니다.
B : 한국 육군을 좀 더 잘 이해하는 데 도움이 되었으면 좋겠습니다.
A : 당신의 친절한 설명에 진심으로 감사드립니다.
B : 자, 이제 전방사단을 방문하겠습니다.
A : 사단의 어디를 방문하실 계획입니까?
B : 우리는 관측소, 일반전초를 방문하고 나서, 장애물과 지뢰지대를 보실 수 있습니다. 또한, 북한군이 굴착한 대남침투용 땅굴도 보시게 됩니다.

1 Observation Post(OP)

A : Now we can see the border fence, right?
B : Yes. This is the northernmost OP on Hill 791.
A : Are there any unusual enemy activities these days?
B : Yes, there have been frequent vehicle movements at night, and strange noises were detected during the daytime. We often observe the soldiers' providing support activities for farming, but those are not unusual.

== Related Expressions ==

- key terrain
- booby trap
- to clear fields of fire
- hardened artillery site
- military demarcation line(MDL)
- demilitarized zone(DMZ)

1 관측소

A : 아, 철책이 보입니다. 그렇죠?
B : 예. 이곳은 791고지에 위치한 최북단 관측소입니다.
A : 최근 적 활동 중 특이한 사항을 관측한 바 있습니까?
B : 예, 통상적인 빈번한 야간 차량 이동과 주간의 특이 소음 등이 탐지됩니다. 간혹 군인들의 영농 지원 활동이 관측되는데 특이한 활동은 아닙니다.

== 관련 표현 ==

- 중요 지형물
- 부비트랩
- 사계청소를 실시하다
- 갱도 포병 진지
- 군사분계선
- 비무장 지대

2 General Outpost(GOP)

A : What is the GOP?
B : It is the 000 security troop.
A : What is the mission of the GOP?
B : First, it is to give warning of the enemy's attack as early as possible. Second, to deceive the enemy and to force its early deployment. Third, to delay the enemy's attack.
A : What are the considerations when employing the GOP unit?
B : We employ it by considering the METT+TC.
A : How does the GOP unit perform its mission?
B : It resorts to deceptional measures, as necessary, it may execute close combat, but it avoids the decisive battle.

== Related Expressions ==

- mission, enemy, terrain and weather, troop+time avaliable, civil affairs(METT+ TC)
- to canalize the enemy into a desired area
- situation report(SITREP)
- to reinforce it with tanks
- to delay the enemy advance as long as possible

2 일반전초

A : 일반전초(GOP)란 무엇입니까?
B : 일반전초는 000의 경계부대입니다.
A : 일반전초의 임무는 무엇입니까?
B : 첫째, 적의 공격을 조기에 경고하고, 둘째, 적을 기만하고 조기 전개를 강요하며 셋째, 적의 공격을 지연하는 것입니다.
A : 일반전초 운용시 고려 요소는 무엇입니까?
B : METT+TC 요소를 고려합니다.
A : 일반전초는 어떻게 임무를 수행합니까?
B : 주로 기만방책을 이용하며 필요시 근접전투를 실시하나, 결정적인 전투는 회피합니다.

== 관련 표현 ===

- 임무, 적 상황, 지형 및 기상, 가용부대 및 시간, 민간요소
- 의도한 지역으로 적을 유도하다
- 상황보고
- 전차로 증강하다
- 적의 전진을 최대한 지연시키다

3 Combat Outpost(COP)

A : What is the COP?
B : It is a security element deployed in front of the main body.
A : What is the size of it?
B : We employ a reinforced 000 up to a 000 unit in front of the battalion's main battle position.
A : What is the principal mission of the unit?
B : In order to provide sufficient warning time to the main body, it observes the enemy's likely avenue of approach along road No. 7 and Hill 298.

== Related Expressions ==

- forward edge of the battle area (FEBA)
- night vision device (NVD)
- thermal observation device (TOD)
- crew-served weapon
- to prepare defensive positions.

3 전투전초

A : 전투전초란 무엇입니까?
B : 전투전초는 주 방어 부대의 경계부대입니다.
A : 그러면 그 규모는 어느 정도입니까?
B : 대대의 주 전투진지 정면에 증강된 000로부터 000규모까지 운용합니다.
A : 전투전초의 주요임무는 무엇입니까?
B : 본대에 충분한 경고 시간을 제공하기 위하여 전투전초는 7번 도로와 298고지를 연하는 적의 예상 접근로를 관측하는 것입니다.

== 관련 표현 ==

- 전투지역 전단
- 야시장비
- 열상 관측 장비
- 공용화기
- 방어 진지를 준비하다

4 FEBA

A : What is the FEBA?
B : It is the front end of the major ground combat units' deployment and the rear area where the covering and screening force are operating.
A : What is the purpose of establishing the FEBA?
B : It is designated to coordinate the fire support, the positioning forces, or the maneuvering of units.
A : When is the FEBA finally established?
B : It is finalized when the forward platoons are deployed on the forward defense area.

== Related Expressions ==

- fire support coordination line(FSCL)
- employment of listening posts
- mine field
- planned targets by direct support artillery
- coordination point
- overhead drops

4 전투지역전단

A : 전투지역전단이란 무엇입니까?
B : 전투지역전단이란 주요 지상 전투부대가 전개하는 전방 끝 선이며 엄호부대와 차장부대가 작전하고 있는 후면입니다.
A : 전투지역전단의 설치 목적은 무엇입니까?
B : 설치 목적은 화력지원과 부대 배치 및 기동을 협조하기 위하여 지정됩니다.
A : 언제 전투지역전단은 최종 형성됩니까?
B : 전투지역전단은 전방 방어지역내의 지상에 전방 소대가 전개되었을 때 최종적으로 이루어집니다.

== 관련 표현 ==

- 화력지원협조선
- 청음초 운용
- 지뢰지대
- 직접지원 포병에 의한 계획된 표적
- 협조점
- 낙석

5 Invasion Tunnels

A : How many tunnels have been found in the DMZ area?
B : There have been four, so far. It was our Division that found the latest one.
A : When did you find it?
B : The reconnaissance unit located it on the 3rd March 1990. One regiment with light combat vehicles can pass through it per hour.
A : I'd like to look at it
B : Look at the digging traces. They are toward the south. You can clearly see that the tunnel was dug from the north to the south.

== Related Expressions ==

- to dig a borehole for tunnel detection.
- to be used as a route for surprise attacks or infiltration
- Defectors testified that they had worked on the tunnels.

5 땅굴 소개

A : DMZ 지역에서 몇 개의 땅굴을 발견하였습니까?
B : 현재까지 4개이며, 우리 사단이 4번째 땅굴을 발견하였습니다.
A : 이 땅굴은 언제 발견하였습니까?
B : 1990년 3월 3일 우리 수색대가 발견하였으며, 시간당 소형 전투차량을 장비한 1개 연대가 통과할 수 있습니다.
A : 현장을 보고 싶군요.
B : 저 굴착 흔적을 보십시오. 북쪽에서 남쪽으로 향하고 있습니다. 이를 보면 이 땅굴은 북한 측에서 굴착 했다는 것을 확실히 알 수 있습니다.

== 관련 표현 ==

- 땅굴 탐지용 시추공을 뚫다
- 기습 공격 혹은 침투 통로로 이용되다
- 귀순자들이 땅굴 작업을 했다는 증언을 했다.

Unit 6. Conference and Pretocol

1 Meeting for training coordination

A : The focus of this training is to cheek the validity of multiple counterattack operation in the direction from Yukdanri to Pyeongyang, one of our corps' contingency plans. What are your thoughts about enemy force size in the penetration?

B : Considering the terrain and current enemy disposition, we estimate that it would be one tank brigade and two mechanized brigades of the enemy's mechanized corps.

A : Make sure that the control plan will be made in ways that can help us to compare the established data base and the results of the multiple counter attack operations.

== Related Expressions ==

- multiple counter attack operations.
- I'd like to welcome all of you to the meeting.
- Thank you for coming to ~.
- Thank you for taking time out of your busy schedule.
- I'd like to announce the meeting open.
- I'd like to declare the conference open.
- It is a great honor and privilege to ~.

Unit 6. 회의 및 의전

1 훈련 협조회의

A : 금번 훈련의 중점은 군단의 우발 계획 중 하나인 육단리에서 평양 방향의 촌단작전의 타당성을 검토하는 것이다. 이때 돌파구 내의 적 규모는 어느 정도 되리라 판단하시오.

B : 지형과 현 적부대 배치를 고려할 때, 적 기계화 군단의 1개 전차 여단과 2개 기계화 여단으로 판단합니다.

A : 이미 구축된 데이터 베이스와 아군의 촌단작전 결과가 검토될 수 있도록 통제 계획을 수립하시오.

== 관련 표현 ==

- 촌단작전
- 회의에 참석한 여러분 모두를 진심으로 환영합니다.
- ~에 참석해 주신 데 대해 감사 드립니다.
- 바쁘신 중에도 참석해 주셔서 감사합니다.
- 이제부터 회의를 시작하도록 하겠습니다.
- 회의의 개회를 선언합니다.
- ~ 할 수 있게 되어 큰 영광입니다.

2 Meeting for combat coordination

A : Could you explain the division's current situation?

B : The attack is being delayed because of the enemy's unexpected resistance, downtown in Gaeseong city and the limitation of other possible routes to bypass the city. Therefore, we estimate the best course of action to secure the Phase Line "Seung Ri" is to transfer priority use of the Road No. 87 to our division

A : Then, I will recommend that the HQ transfer the priority use of the road No. 87 to "A" division as of 0200 hours on 14 July.

B : When we advance along road No. 87, we may face difficulties in the river crossing due to the possible counter-attack by the enemy's Mech. brigade located to the east of Mt. Moak. Therefore, we need to take additional actions.

== Related Expressions ==

- The meeting consists of two sessions.
- We are gathered here to discuss ~.
- The purpose of the meeting today is to discuss ~.
- The agenda of today's meeting is shown on the chart.
- Because of time constraint, each speaker will have just 10 minutes.

2 전투 협조회의

A : 사단의 현 상황은 어떻습니까?
B : 개성 시가지에서의 적의 예상치 못한 저항과 우회로 제한에 따라 공격이 지체되고 있습니다. 따라서 87번 도로 사용 우선권을 아 사단으로 전환하여 주시면, 부여된 시간 내에 통제선 "승리"를 확보할 수 있다고 판단됩니다.
A : 그렇다면, 7.14, 02:00부로 87번 도로 사용 우선권을 A 사단에 부여토록 건의하겠습니다.
B : 87번 도로를 따라 모악산 동측에 위치한 적 기계화 여단의 역습에 따른 하천 장애물 극복에 어려움이 있으리라 생각됨으로 추가적인 조치가 필요합니다.

== 관련 표현 ==

- 회의는 1, 2부로 구성되어 있습니다.
- 우리는 ~을 토의하기 위해 이곳에 모였습니다.
- 오늘 회의의 목적은 ~을 토의하는 것입니다.
- 오늘 회의에서 토의할 사항은 차트에서 보시는 바와 같습니다.
- 시간 관계상 각 발표자의 발표시간은 10분만 주어집니다.

3 Meeting for CSS coordination

A : Do you have any logistic problems in supporting the current OPLAN?
B : We estimate we may face difficulties in ammunition supply if we are attacked by the enemy's aviation and missiles.
A : In particular, Hill Galma would be a major weak point for ammunition supply, wouldn't it?
B : Yes. So we have a plan to use a mobile ammunition supply point at the rear of Hill Galma.
A : Let me add one more option. I think we can adjust the CSR in response to the enemy's resistance.
B : Besides, the special purpose ammunition needed in the initial phase should be accumulated in the front area as soon as possible.

== Related Expressions ==

- I'd like to invite COL Jeon to make a comment.
- Any other idea or different opinions?
- Let's move onto the next issue.
- Now, we'd like to take questions from the floor.
- I'd like to ask Maj Kim to explain it in more detail.
- I have a question for Maj Lee.
- Does that answer your question?

3 작전지속지원 협조회의

A : 군수지원 분야에서 현 작전계획 지원에 문제는 없습니까?
B : 만일 적 공중, 미사일 공격을 받는다면 탄약 보급에 어려움이 따르리라 판단됩니다.
A : "갈마"고개가 탄약 보급에 커다란 취약 요소가 될 수 있겠지요?
B : 예. 그래서 우리는 갈마 고개 후사면에 탄약 분배소를 운용할 계획입니다.
A : 한가지 더 추가한다면, 적의 저항에 따라 통제 보급률을 신축성 있게 조정하는 방안도 있습니다.
B : 그 밖에 초전 임무 수행에 필요한 특수 목적탄은 최대한 추진 비축토록 해야 되겠습니다.

== 관련 표현 ==

- 전 대령님 견해를 밝혀 주시겠습니까?
- 다른 의견 있으시면 말씀해 주십시오.
- 다음 문제를 토의하겠습니다.
- 방청석에서 질문을 받겠습니다.
- 김 소령님, 그것에 대해 좀 더 자세히 설명해 주시겠습니까?
- 이 소령님께 질문 드리겠습니다.
- 질문에 답이 되었습니까?

4 After Action Review

A : I'd like you to make a comment on the FTX of the mechanized division.

B : As Mechanized Division Commander, I learned a lot of aspects different from the simulation training.

A : Could you tell us in detail?

B : Okay. First, we considered only tactical factors such as unit maneuver and combat service support in the CBT training, but in the FTX, I think we should pay more attention to junior officers training in obeying directions in the field. Second, I should have made decisions in consideration of the biorhythms of the crew for day and night. Third, we should pay more attention to avoid civil damage to civilians.

== Related Expressions ==

- I agree with you on the whole, but ~.
- I don't accept that.
- Fine, that's settled then.
- I'm not in a position to ~
- One way to solve this problem is to ~
- Thank you for your excellent presentation.
- I'd like to close this meeting by thanking all the speakers and participants.

4 훈련 후 강평

A : 기계화 사단 실병기동훈련을 실시한 소감을 들어보고 싶습니다.
B : 기계화 사단장으로서 그동안 시뮬레이션을 통한 훈련시와는 다른 면을 체험하였습니다.
A : 구체적으로 말씀하신다면……?
B : 먼저, CBT 훈련시에는 부대 기동과 작전지속지원 등 전술적인 면만 고민하였습니다. 그러나 실병기동시에는 방향유지를 위한 초급 장교들의 훈련이 절실하였습니다. 둘째는, 생체리듬, 주 · 야 시간대별 조종수를 포함한 팀 능력 등을 고려한 전술결심을 했어야만 했습니다. 셋째는, 대민 피해 방지에 좀 더 주의를 기울여야 한다는 것입니다.

== 관련 표현 ==

- 대체로 찬성입니다, 그러나 ~
- 그것을 받아들일 수 없습니다.
- 좋습니다. 그것으로 결정했습니다.
- 저는 ~ 할 입장(위치)에 있지 않습니다.
- 이 문제를 풀기 위한 하나의 방법은 ~ 하는 것입니다.
- 훌륭한 발표에 감사의 말씀을 드립니다.
- 모든 발표자와 참석자들께 감사의 말씀을 드리며 회의를 마치겠습니다.

5 Armed Forces' Day

A : Today, we celebrate the 75th anniversary of the ROK Armed Forces' Day.
B : Why did they designate the 1st October as the Armed Forces' Day?
A : Because it is the day we advanced through the 38th parallel line after recovering Seoul in the Korean War.
B : Where are the seats for foreign officers?
A : They are on the right side of the VIP seats. I'll guide you.
B : How long will the ceremony last?
A : It will take about an hour. It begins with an honor to the President. A twenty- one gun salute will be fired at this point.
B : What is next?
A : Then, honors to the nation, the President's address, inspection of troops, and a parade will follow.

5 국군의 날

A : 오늘은 창군 제75주년 대한민국 국군의 날입니다.
B : 왜 10월 1일로 정하였습니까?
A : 한국전쟁 시 서울 수복 후 최초로 38선을 통과하여 북진한 날이기 때문입니다.
B : 외국군 장교석은 어디죠?
A : 귀빈석 우측입니다. 안내하겠습니다.
B : 행사 시간은 어느 정도 소요됩니까?
A : 약 1시간이며, 대통령께 대한 경례로부터 시작됩니다. 이때 예포가 21발 발사됩니다.
B : 그다음 행사는?
A : 국기에 대한 경례, 대통령 치사, 열병 및 분열 등입니다.

6 Ceremony in the National Cemetery

A : The Division Commander will participate in a ceremony honoring the deceased soldiers at 10:00 A.M. on the 6th of June.
B : Could you tell me the procedure of the ceremony?
A : It will start with laying flowers followed by the moment of silence. The whole ceremony will take about five minute.
B : Who are resting at this part of cemetery?
A : The Korean War veterans.
B : What are the major events of the ceremony?
A : During the moment of silence, requiem and condolence fire will be conducted.

6 국립 현충원 참배

A : 사단장께서 6월6일 오전 10:00에 국립현충원 현충탑을 참배할 계획입니다.
B : 참배 절차는 어떻게 됩니까?
A : 도착 후 헌화, 묵념 순으로 행사는 약 5분 소요됩니다.
B : 참배할 묘역은 어떤 분들이 있습니까?
A : 한국전 참전용사들입니다.
B : 참배 간 유념해야 할 사항은 무엇입니까?
A : 묵념 간 진혼곡과 조총 사격이 있을 예정입니다.

Unit 7. Staff Responsibilities and Duties

1 Personnel

(1) Personnel Management

A : G-1, what is the current personnel strength of the corps in compared to the authorized strength?
B : It is 98% of officers and 94% of enlisted men as of the 31st of August.
A : Are there any persons eligible for corps G-4 among the regiment commanders?
B : Yes, sir! There are two colonels at the 16th and the 24th ID who are supposed to complete their assignments at the end of January.
A : Recommend one after considering career, ability, and integrity.

== Related Expressions ==

- to be assigned to ~
- designation, recommendation
- promotion board
- officer evaluation report(OER)
- permanent change of station(PCS)

Unit 7. 참모업무

1 인사

(1) 인사 관리

A : 인사참모, 현재 군단의 인가 대비 병력 현황은 어떻게 되는가?
B : 8월 31일 현재 장교 98%, 병 94%입니다.
A : 연대장들 중 군단사령부 군수참모로 적절한 대상자가 있는가?
B : 예, 16사단, 24사단에 각각 1명씩 1월에 연대장 임기를 마치는 장교가 있습니다.
A : 군수참모 추천 시 관련 업무 경력과 능력, 성실도를 주로 고려하시오.

== 관련 표현 ==

- ~에 보직되다
- 지명, 건의
- 진급 심사 위원회
- 장교 근무 평정
- 전속

(2) Safety Service

A : It is expected that this year will be hotter and more humid than last year. Therefore, you should develop a plan to prevent various accidents which may occur during training, and check the MEDEVAC system.

B : Education on how to prevent food poisoning, sun stroke and heat stroke has already been conducted for ten battalions which are to participate in the exercise next month. In the past, dysentery, a summer epidemic, occurred in this area, but recently malaria has also been found. We coordinate closely with local health and medical agencies in order to prevent disease.

A : Two battalions are scheduled to go out for mountain training next week, so the supporting aviation unit should be ready for emergency medical evacuation.

== Related Expressions ==

- treatment of mass casualties
- temporary burial
- to carry first aid medicine
- blood bank
- medical evacuation(MEDEVAC)

(2) 안전 근무

A : 올해는 보다 무더운 날씨가 예상되는데, 훈련 간 발생할 수 있는 각종 사고 예방을 위한 계획과 적시적인 구조활동 체계를 발전시킬 수 있도록 하시오.

B : 다음 달부터 훈련할 10개 대대에 대한 식중독 예방, 일사 · 열사병 예방을 위한 교육은 이미 실시되었습니다. 과거에 지역 내에서 하절기 전염병 이질이 발생한 적이 있으며, 최근에는 말라리아도 발생하였습니다. 그래서 우리는 질병 예방을 위하여 지역 보건소, 의료기관과 긴밀한 협조를 하고 있습니다.

A : 다음 주 산악훈련을 할 부대가 2개 대대 있는데 군단 항공대는 응급환자 후송을 위한 준비를 갖추도록 하시오.

== **관련 표현** ==

- 대량 전사상자 처리
- 가매장
- 응급 약품 휴대
- 혈액 은행
- 의무 후송

(3) Civil-Military Friendship Activities

A : The annual civil-military friendship activities will be held on the 6th and 7th October. We'll be joined by 50 residents in Doowon Myeon and about 30 soldiers from the 17th U.S. Aviation Brigade for the activities.

B : What are the major events?

A : On the first day, we are going to have unit orientation, Corps Commander's meeting with the local representatives, and the Army Band's performance. On the second day, we'll have a mountain climbing and environment protection activity.

B : We need to have common activities more frequently rather than having just an annual activity. For example, I would recommend playing soccer with the command staff, local residents, and US soldiers together once a week at the command field, and support of the English class in the middle school by US soldiers.

== Related Expressions ==

- to promote mutual understanding
- to resolve the environmental issues on a mutual basis
- timely disaster relief
- preventive activities against disaster

(3) 민군 친선 활동

A : 이번 연례 민군 친선행사는 10월 6일, 7일 양일로 정하였습니다. 대상은 사령부 간부들과 두원면 주민 50여명이며, 17 비행단에 근무하는 미군 약 30명이 함께할 예정입니다.

B : 주요 행사 내용은?

A : 첫 날은 부대 소개 및 군단장님과 주민 대표 간담회, 군악 연주회 등이 계획되어 있습니다. 둘째 날은 등산 및 자연보호 활동을 준비하였습니다.

B : 한 차례의 연례행사보다, 자주 함께 하는 활동들이 요구되는데, 예를 든다면, 사령부 연병장에서 주 1회 정도 사령부 간부 · 지역 주민 · 미군이 함께하는 축구, 미군의 인접 중학교 영어 수업지원 등을 들 수 있을 것이오.

== **관련 표현** ==

- 상호 이해 증진
- 환경문제 공동해결
- 적시적인 재난구조
- 재해 예방 활동

2 Intelligence

(1) Recent Enemy Activities

A : We observed frequent vehicle movements on Road No. 237 from Geojin to Jangjin on the 17th and 18th of August. A similar situation was observed in July last year. We estimate that this is a preparation for the Annual National Inspection.

B : Did you identify any movements of the mechanized unit located in the rear of Tonggu?

A : Yes. We estimate only the command group took part in the exercise with twenty armored vehicles moving from Tonggu to the front area.

== Related Expressions ==

- farming support
- deployable intelligence support element
- second echelon
- Daily Intelligence Summary (DISUM)
- command group

2 정보

(1) 최근 적 활동

A : 8월 17일, 18일 양일간에 걸쳐 거진에서 장진에 이르는 237번 국도상의 야간 차량이동이 빈번해 졌습니다. 이와 유사한 상황은 작년 7월말에도 포착되었던 것으로 매년 실시되는 국가지도 검열로 판단됩니다.
B : 통구 후방에 위치한 것으로 식별된 기계화 부대의 움직임은 없는가?
A : 예. 통구 지역에서 전방으로 이동한 부대는 약 20대의 장갑차량으로 지휘부만 훈련에 참여하는 것으로 판단됩니다.

== 관련 표현 ==

- 영농지원
- 기동 정보 지원반
- 제 2제대
- 일일 정보 요약
- 지휘부

(2) Psychological Operation

A : You have some difficulties performing the psychological operations these days, don't you?

B : Yes, now that the North Korean side has stopped provoking broadcasts, we are planning appropriate actions.

A : In order to cope with the current situations on the Korean Peninsula, you should analyze the North Korean broadcasts very carefully.

B : Recently, North Korean broadcasts mainly cover the achievements of Kim Jong-Il.

A : We need to employ the morale building teams and the entertainment teams along the GP and GOP area. We also should develop the additional psychological measures.

== Related Expressions ==

- leaflet drop operations
- loudspeaker broadcast
- Combined Psychological Operations Task Force(COPTF)
- leaflet bomb
- consolation broadcast

(2) 심리전

A : 요즘은 심리전 수행에 어려움이 많은 것 같은데?
B : 예, 북한측에서 이전의 적대적인 비방방송이 사라졌으므로 우리도 상응하는 조치를 구상중에 있습니다.
A : 최근 남북관계 변화상황에 대비하기 위하여 북한군 대남방송 등을 잘 분석하여야 합니다.
B : 최근 전선지역 북한군 방송은 김정일의 치적에 바탕을 둔 방송이 많이 청취되고 있습니다.
A : 그럼 우리도 GP와 GOP 지역에 문선대 등을 보내어 자체 위문활동 실시하고, 기타 대적 심리전 효과를 얻는 방안을 발전시키시오.

== 관련 표현 ==

- 전단 살포 작전
- 확성기 방송
- 연합 심리전 사령부
- 지폭탄
- 선무 방송

(3) Operations Security

A : We'll inspect the operations security status of the subordinate units for three days this week.
B : What area are you going to focus on?
A : The control measures used in accessing each unit's computer terminals and observance of the related regulation will be carefully inspected.
B : In addition to that, please check the visitor's entry procedures to the garrison and the use of security devices.
A : Okay. We'll inspect the security devices with the assistance from the defense security unit.

== Related Expressions ==

- civil-defense exercise
- restricted area
- password
- secure phone
- unannounced security inspections
- security clearance

(3) 작전 보안

A : 이번주 3일간 각 예하부대의 작전보안 상태를 점검토록 하겠습니다.
B : 어디에 중점을 두고 실시합니까?
A : 각 부대의 전산실 출입통제실태, 관련 규정준수 여부 등에 중점을 두고 실시할 예정입니다.
B : 거기에 추가하여 면회자 부대 출입 절차 준수여부 및 보안장비 활용실태도 파악토록 하시오.
A : 알겠습니다. 보안장비는 지원 방첩부대의 협조하에 점검토록 하겠습니다.

== 관련 표현 ==

- 민방위 훈련
- 출입통제 구역
- 암구호
- 비화 전화기
- 불시 보안점검
- 신원조회

3 Operations

(1) Security Mission

A : According to the recent check-up of the security duty along the fence line, there are some points to be improved.

B : Are they related to training and economy of force for the GOP unit?

A : Yes, that's right. One battalion is fully responsible for 4km of frontline and three companies are deployed for security duty. We could take one company out of the security duty if we employ surveillance equipment well in some parts of the area. We can transfer the company to a reserve, rotate the mission in every two or three months, and provide the reserve company with the sufficient time to train the soldiers.

== Related Expressions ==

- initial response cell
- employ more guard posts
- local security forces
- intensive employment of night vision devices
- chain of command

3 작전

(1) 경계 근무

A : 전방 철책지역 경계근무 실태 점검결과 몇가지 개선되어야 할 부분들이 있었습니다.
B : GOP 담당 부대의 훈련문제와 병력절약 운용 사항입니까?
A : 예, 그렇습니다. 1개 대대가 4km의 정면을 전담하고 있으며, 3개 중대가 전방에 전개되어 경계근무 중입니다. 일부 지역은 감시장비 등을 잘 운용하면 1개 중대 정도는 절약할 수 있습니다. 이를 예비로 하여 2-3개월 간격으로 교대를 실시하여, 병사 훈련을 위한 충분한 시간을 제공할 수 있습니다.

== **관련 표현** ==

- 초기대응반
- 증가초소 운용
- 국지 경계 부대
- 감시장비 집중 운용
- 지휘 계통

(2) Operations Plan

A : What is the emphasis of this OPLAN discussion?

B : The emphasis of the discussion is how to develop the COAs to force the enemy's second echelon to commit itself earlier.

A : Do we have sufficient fire assets to ensure the friendly armor brigade's maneuver along FEBA and to block the enemy advance and destroy them?

B : The road nets in column are well developed, but the road net in row is not well developed. The fire support from ten 155mm field artillery battalions and one attack helicopter battalion under OPCON is available. We should be prepared to request naval gun fires from the 5th Fleet because the enemy may maneuver along the developed seashore road to the south.

== Related Expressions ==

- course of action(COA)
- offensive employment of armor units
- close coordination with maneuver and fire
- road condition
- covering force

(2) 작전 계획

A : 이번 작전계획 토의의 중점은 무엇입니까?
B : 작계토의 중점은 적 제 2제대의 조기 투입을 강요하기 위한 방책을 발전시키는 것입니다.
A : 전투지역전단에서의 아 기갑여단의 기동보장, 적 전진의 저지 및 격멸을 위한 화력자산은 충분합니까?
B : 종적 도로망은 잘 발달되어 있으나 횡적 도로망이 제한됩니다. 화력은 155밀리 10개 포병대대와 작전 통제된 공격헬기 1개 대대의 지원이 가능합니다. 또한 적이 남쪽으로 발달된 해안도로를 따라 기동 가능하기 때문에 5함대사 함포지원도 요청할 준비를 하겠습니다.

== **관련 표현** ==

- 방책
- 기갑 부대의 공세적 운용
- 기동과 화력의 긴밀한 협조
- 도로상태
- 엄호 부대

(3) Obstacle Plan

A : Have you been informed of the obstacle plan from the higher command?
B : Yes, sir. We received the plan on the 27th August.
A : In accordance with the plan, develop our own plan in detail.
B : We will check the status of the current obstacle material management and request the additional supplies.
A : Based on the obstacle plan, make an exact estimate of the amount of obstacle materials, reception, storage, transportation routes, and distance.

== Related Expressions ==

- obstacle plan
- obstacle and denial operation
- road crater
- storage point

(3) 장애물 계획

A : 상급 부대로부터 장애물계획을 통보 받았는가?
B : 예. 8월 27일에 받았습니다.
A : 상급부대 계획에 따라 우리의 세부계획을 수립하시오.
B : 예, 저희는 기존 장애물자재 관리현황을 확인하고 추가 보급을 건의하겠습니다.
A : 장애물계획에 따른 소요자재 보급 건의, 수령, 저장, 수송경로 및 거리를 판단하시오.

== 관련 표현 ==

- 장애물 계획
- 장애물 및 거부 작전
- 도로 대화구
- 저장소

4 Logistics

(1) Ammunition Support

A : Do you have any shipping plans for special ammunition like MLRS?
B : Yes, we have. The current amount of MLRS ammunition is enough to support the operations for one months.
A : Aren't the supply routes limited?
B : If we control the residents as planned, we think we can support mission successfully.
A : I think preparing alternative routes is also very important

== Related Expressions ==

- basic load of ammunition
- distribution point
- weapon storage
- available supply rate(ASR)

4 군수

(1) 탄약 지원

A : 다련장 포탄 등 특수탄 적송계획은 있습니까?
B : 예. 세부 적송계획은 준비되어 있으며, 다련장은 현 보유탄으로 1개월 간은 작전이 가능합니다.
A : 보급로가 제한되지 않겠는가?
B : 계획대로 주민 통제계획이 제대로 시행된다면, 우리는 지원임무를 성공적으로 수행할 수 있다고 봅니다.
A : 예비 보급로를 준비하는 것도 또한 매우 중요하다고 생각됩니다.

== 관련 표현 ==

- 탄약 기본 휴대량
- 분배소
- 무기고
- 가용 보급률

(2) Transportation

A : Did you coordinate with the Air Force and Navy units located in the area of operations about the route priority

B : The shipping for the Navy will be made by sea routes except the Class I, and the transportation for the Air Force will be done by air.

A : One armored brigade of the Corps is planned to be moved to the Corps AOR by rail lines. Are there any problems?

B : There will be no problem because the 27th Air Transportation Wing will complete the materiel transportation by D-1.

A : Make sure that the bridges are strong enough for tanks to cross.

== Related Expressions ==

- area of responsibility(AOR)
- open storage
- general data of the railroads
- unloading equipment
- highway regulating point(HRP)
- turnaround time

(2) 수송

A : 작전지역 내에 위치한 공군부대 및 해군부대와의 육로 사용 우선권을 협조하였는가?
B : 해군은 1종 외에는 전량 해상을 통하여 이루어지고, 공군은 항로를 이용, 수송토록 되어 있습니다.
A : 군단의 1개 기갑 여단이 철로를 이용하여 군단 책임 지역으로 이동하도록 계획되어 있다. 문제가 있는가?
B : 제27 수송비행단이 D-1일까지 물자수송을 완료하기 때문에 문제가 없습니다.
A : 교량들이 전차 통과에 없는지 확인하시오.

== 관련 표현 ==

- 책임지역
- 야적장
- 철로 일반제원
- 하역 시설 장비
- 육로 조정 통제소
- 왕복수송 소요시간

5 Communications

A : Are we on the right schedule for doing preventive maintenance of equipment and LAN system for office automation?
B : Yes, we are regularly doing it once a month.
A : We couldn't gain access to the signal intelligence. Could you tell me why?
B : Yes, that's due to a circuit outage between the Field Army HQ and the corps HQ.
A : Do we have any security problems with our computer network?
B : We are taking all the security measures including using passwords and security equipment, we immediately inform the related offices and get some help.

== Related Expressions ==

- local area network(LAN)
- terminal
- signal intelligence(SIGINT)
- office automation
- virus check
- to encode data

5 통신

A : 사무자동화용 장비 및 LAN 시스템의 예방정비는 계획대로 실시하고 있습니까?
B : 예, 월 1회씩 정기적으로 실시하고 있습니다.
A : 신호 정보자료를 열어볼 수 없는데 이유가 있습니까?
B : 예, 군사령부와 군단 간의 전산망 회선 불량 때문입니다.
A : 우리 부대의 전산 보안상의 문제는 없습니까?
B : 패스워드, 보안장비 운용 등 제반 보안조치들을 취하고 있습니다. 만약, 보안 장비상의 문제가 발생하면, 즉각 관련 부서에 통보하여 조치를 받고 있습니다.

== 관련 표현 ==

- 근거리 통신망
- 단말기
- 신호 정보
- 사무자동화
- 바이러스 점검
- 데이터를 암호화하다

6 Public Affairs

A : There was a report on the friendship activity between our Corps and the US Army in the newspaper Chosun Ilbo.
B : Can you tell me more specifically?
A : It's about the recent unit ceremony and our support activity for residents.
B : What is the source of the news?
A : The residents in the village.
B : Good. Be sure to provide a formal press release to the media.

== Related Expressions ==

- Public Affairs Office(PAO)
- breaking news
- correspondent
- mobile public affairs team(MPAT)
- press conference

6 공보

A : 조선일보에 우리 군단과 미군과의 친선 활동이 보도되었습니다.
B : 구체적으로 어떤 내용입니까?
A : 최근 부대 행사와 지역주민 지원 활동에 관한 것입니다.
B : 자료제공은 누가 하였습니까?
A : 지역 주민입니다.
B : 좋습니다. 사령부의 공식 보도자료를 언론에 제공하도록 하시오.

== 관련 표현 ===

- 공보실
- 긴급 뉴스
- 특파원
- 이동 공보팀
- 기자회견

7 Management Affairs

A : Are you controlling the meetings and reports?
B : Yes, sir. I have been controling unnecessary meetings and reports in order to ensure smooth subordinate unit activities.
A : All right. On the top of that, conduct the budget operation analysis for the 2nd quarter and be prepared to do the quarterly closing of books.
B : I will do the assignments and report back to you, sir.

== Related Expressions ==

- job analysis
- apportionment
- funds execution
- allowances
- fiscal year(FY)

7 관리

A : 회의 및 보고 통제를 실시하고 있는가?
B : 예. 그 동안 예하부대 활동 보장을 위하여 불필요한 회의 및 보고사항은 통제하여 왔습니다.
A : 추가적으로 2/4분기 자금 운영분석을 실시하여 분기 결산을 준비토록 하시오.
B : 임무 수행 후 보고드리겠습니다.

== **관련 표현** ==

- 직무 분석
- 자금 할당
- 자금 집행
- 수당
- 회계 연도

Unit 8. Unit Management

Maj Lee explains to Maj William how to manage the unit.

A : What are the major factors maintaining the unit combat power?
B : In addition to training and education, we also consider unit management very important.
A : What are the key points in unit management?
B : Personnel management is one of the most important things. Additionally, the management of facilities and equipment is involved.
A : What is the best way for successful unit management?
B : The key to the success is to unite closely together.

== Related Expressions ==

- Army Regulations(AR)
- resource management
- to maintain military discipline
- unit strength status

Unit 8. 부대관리

이 소령은 윌리엄 소령에게 부대 관리 요청을 설명한다.

A : 부대 전투력을 유지하기 위한 요소는 뭐라고 생각하십니까?
B : 교육훈련과 더불어 부대관리가 중요합니다.
A : 부대관리에서 무엇이 가장 중요한 사항입니까?
B : 전체에서 가장 중요한 요소 중의 하나는 병력관리입니다. 추가적으로 기타 시설 및 장비관리가 포함될 수 있습니다.
A : 부대관리의 성공을 위해 가장 중요한 것은 무엇입니까?
B : 성공의 관건은 전 부대원이 단결하는 것입니다.

== 관련 표현 ==

- 육군 규정
- 자원 관리
- 군 기강을 유지하다
- 부대 병력 현황

1 Daily Schedule

A : What time does the work start?
B : It starts at 0800 hours.
A : What is the difference between the ROK Army and the US Army in terms of the daily schedule?
B : Since we have conscription system for the Army, all the soldiers stay in barracks and have roll call in the mornings and in the evenings. After work, they have maintenance time, dinner, and free time.

== Related Expressions ==

- command sergeant major(CSM)
- to have four days off as a reward for being a distinguished marksman
- to convert to the winter time schedule
- to make good use of free time

1 일과표

A : 몇시에 일과가 시작됩니까?
B : 08:00에 시작됩니다.
A : 미군과 한국군의 일과의 큰 차이가 무엇입니까?
B : 한국군은 징병제라서 병사들이 영내에 거주하여 아침, 저녁으로 점호를 취합니다. 일과 후에는 정비 시간, 저녁 식사 그리고 자유시간이 있습니다.

== 관련 표현 ==

- 주임원사
- 전투사격 우수자로 3박 4일 특박을 부여받다
- 동계 일과표로 변경하다
- 자유시간을 적절히 활용하다

2 Accident Prevention

A : What are the roles of the officers and NCOs in preventing accidents?
B : First, they need to lead by example to implement the safety regulations and policies. Second, they should conduct safety education and recognition activities for the soldiers. Finally, complete supervision is required.
A : How do you conduct safety education?
B : First, we conduct danger recognition training prior to unit activites. In addition, we disseminate the relevant accident cases in order to keep up awareness.
A : Have accidents decreased after safety training by the commanders?
B : Yes, the accidents have been decreasing year after year. This is a prerequisite in order to maintain our combat power.

== Related Expressions ==

- Prepare the safety measures
- Analyze and report the causes of the accidents.
- Do check-ups on a regular basis
- major accidents in summer : drowning, killed by the poisonous insects, food poisoning, floods, sunstroke and heatstroke
- major accidents in winter : burns, frostbite, injuries from slipping and falling, carbon monoxide poisoning
- other accidents : vehicles, weapons, and explosives
- Accident Prevention Week

2 사고 예방

A : 사고 예방을 위한 간부의 역할은 무엇입니까?

B : 첫째는 각종 안전규정 이행을 위해 솔선수범해야 합니다. 둘째, 부하들에게 안전교육 계몽활동을 실시합니다. 마지막으로 철저한 확인감독이 요구됩니다.

A : 사고 예방 교육은 어떻게 실시합니까? 먼저 부대 활동 전 위험예지 훈련을 실시합니다. 또한 유사사고 사례 전파로 경각심을 제고시킵니다.

B : 지휘관에 의한 안전교육 후 사고는 많이 감소되었습니까?

A : 예, 매년 안전사고는 감소하고 있습니다. 이것은 바로 군 전투력 유지의 필수 요소입니다.

== 관련 표현 ==

- 사고예방 대책을 수립하라.
- 사고의 원인을 분석하여 보고하라.
- 확인점검을 정기적으로 해라
- 하계 다발사고 : 익사, 교사, 식중독, 홍수 및 붕괴, 일사 또는 열사
- 동계 다발사고 : 화상, 동상, 낙상, 일산화탄소 중독
- 기타 사고 : 차량 안전, 총기 및 폭발물
- 사고 예방 주간

3 Vehicle Maintenance

A : What kind of maintenance cycles do you have for vehicles?
B : Vehicle maintenance is the key to safety, so we do the maintenance on a daily, weekly, monthly, biannually, and yearly basis.
A : How do you conduct the maintenance?
B : Mechanics and drivers do vehicle maintenance under the supervision of the maintenance officer as scheduled.
A : Why is vehicle maintenance so important?
B : Because inappropriate maintenance can cause a lot of problems in combat readiness.

== Related Expressions ==

- types of vehicle : jeep, van, truck
- Keep to the speed limit.
- What kind of the maintenance schedules do you have?
- maintenance schedule
- defensive driving

3 차량정비

A : 차량정비 주기에는 어떤 것이 있습니까?
B : 차량 정비는 안전에 중요 요소이기 때문에, 일일, 주간, 월간, 반년, 연간 정비를 실시하고 있습니다.
A : 정비는 어떻게 합니까?
B : 정비병과 운전병들은 정비관의 감독 하에 계획된 데로 차량을 정비합니다.
A : 차량 정비는 왜 그렇게 중요합니까?
B : 그것은 차량 정비 상태가 나쁘면 부대의 전투 준비 태세에 문제를 야기할 수 있기 때문입니다.

== 관련 표현 ==

- 차량종류(짚, 밴, 트럭)
- 제한 속도를 준수하라.
- 정비의 계획에는 무엇이 있는가?
- 정비 계획표
- 방어운전

4 Activities after Duty Hours

A : What do soldiers usually do after duty hours?
B : After duty hours, soldiers have dinner and enjoy free time.
A : What do the soldiers usually do during their free time?
B : They enjoy club activities, watching TV, and physical activities.
A : Are there any activities that commanding officers encourage?
B : Yes, commanding officers support activities such as surfing the internet, religious activities, and watching movies.

== Related Expressions ==

- club activities
- Get yourself out of a passive life style.
- the operating hour of the PC room
- flag raising ceremony

4 일과 후 활동

A : 일과 후에 병사들은 주로 무슨 활동을 하게 됩니까?
B : 교육훈련을 마친 후 병사들은 저녁을 먹고 나서, 자유시간을 갖습니다.
A : 자유시간에 병사들은 보통 무엇합니까?
B : 그들은 동아리 활동이나 TV 시청, 운동 등을 합니다.
A : 지휘관이 권장하는 활동이 있습니까?
B : 예, 인터넷 서핑, 종교활동, 영화감상 등의 활동을 지원해 주고 있습니다.

== 관련 표현 ==

- 동아리 활동
- 소극적인 생활에서 벗어나라.
- PC방 운영시간
- 국기게양식

5 Sports Activity

(1) Soccer

A : What kind of sports are played on the base?
B : Soccer, basketball and tennis are played on the base.
A : What is the most popular sport?
B : Soccer, I guess. Since World Cup 2002 will be held in Korea and Japan, it's getting more popular.
A : Do you play soccer by team?
B : Yes, 'A' Company is playing soccer against 'B' Company.[3)]

== Related Expressions ==

- to play soccer
- Army Physical Fitness Test(APFT)
- The International Football Federation rules.

3) 육군사관학교 영어과, 『Military English Ⅰ』. p.72.

5 체육 활동

(1) 축구

A : 영내에서는 어떠한 운동을 주로 합니까?
B : 부대 내에서는 축구, 농구, 야구, 테니스 등을 하고 있습니다.
A : 가장 선호하는 종목은 무엇이죠?
B : 저는 축구라고 생각해요. 2002년 월드컵이 한국과 일본에서 개최되어 더욱 인기가 있죠.
A : 부대별 축구경기를 합니까?
B : 예, 최소한 대대별로 1개의 축구장을 보유하고 있습니다.

== 관련 표현 ==

- 축구를 하다
- 육군 체력 검정
- 국제 축구 연맹 규칙

(2) Basketball

A : In the U.S Army, basketball is very popular. Is it popular here in the ROK Army, too?
B : Yes, basketball is getting popular these days among the soldiers.
A : Oh, I see
B : And the professional basketball league makes basktball popular as much as soccer.
A : I can see a basketball court over there.
B : Yes. That's right. Company-size units have at least one basketball court. We can play basket ball even in small areas.

== Related Expressions ==

- basketball backboard
- We have basketball uniforms with the players' numbers on the back
- The ball does bot bounce well.
- to inflate the ball

(2) 농구

A : 미 육군에서는 농구가 매우 인기 있는데, 여기 한국 육군에서도 인기가 있습니까?
B : 예, 요즘 우리 병사들 사이에서도 농구가 인기 있습니다.
A : 아, 그렇군요.
B : 더구나 프로 농구가 생겨서 축구에 버금갈 정도입니다.
A : 저쪽에 농구장이 보이는군요?
B : 예. 맞습니다. 예하부대는 중대급에 1개 정도의 농구장을 보유하고 있습니다. 특히 좁은 공간에서도 쉽게 할 수 있는 운동입니다.

== **관련 표현** ==

- 농구대
- 중대에는 등번호판이 부착된 농구복이 있다.
- 공이 바운드가 잘 되지 않는다.
- 공에 바람을 넣다

(3) Tennis

A : Do the officers and soldiers play the same sports?
B : No. Though they play together, the officers usually play tennis.
A : Why do they play tennis a lot?
B : Because it can be played with a small number of players, and teams can be made up easily.
A : When do you play it most?
B : We normally do so on weekends, national-holidays or during physical training time.

== Related Expressions ==

- to stretch / warmup
- to play every weekend with people who like tennis
- to have a big score gap
- to equalize to tie the score
- In tennis, there is a practice court and a match court.
- to drive, cut, volley, slice, smash, stroke, Job serve

(3) 정구

A : 장교와 병사들은 동일한 체육활동을 합니까?
B : 아닙니다. 같이 하는 경우가 많으나, 간부들은 주로 정구를 많이 합니다.
A : 정구를 많이 하는 이유가 무엇입니까?
B : 적은 인원으로 할 수 있는 운동이라서 쉽게 팀이 구성됩니다.
A : 언제 많이 합니까?
B : 일반적으로 주말, 휴일이나 전투체육시간을 이용합니다.

== 관련 표현 ==

- 준비 운동을 하다.
- 테니스를 좋아하는 사람끼리 주말마다 운동을 한다.
- 점수 차이가 많이 나다.
- 동점을 이루다.
- 정구에는 연습용과 경기용이 있다.
- 드라이브, 컷트, 발리, 슬라이스, 스매싱, 스트로크, 로빙

6 Ammunition Management

A : How do you manage ammunition storage?
B : Ammunition is maintained in a separated facility according to the safety regulations.
A : What are the safety measures?
B : First, we prohibit anyone from carrying anything flammable near the ammunition storage. Second, we set up barriers and barbed-wire to restrict access. Finally, we maintain the security around the clock.
A : What is the most important thing to secure the ammunition storage?
B : We should set up racks in the ammunition storage to prevent rust, and exhaust fans are used yo maintain ventilation.

== Related Expressions ==

- periodic inspection and stock rotation
- to clear the fields of fire, set up firebreaks
- KEEP OFF
- to check by counting
- high explosive dual-purpose(HEDP)

6 탄약 관리

A : 탄약은 어떻게 관리하고 있습니까?
B : 탄약은 별도의 시설인 탄약고에 안전규정에 의거 보관하고 있습니다.
A : 안전대책에는 어떠한 것이 있습니까?
B : 먼저, 탄약고 주변에는 인화물질 휴대를 금하고 있으며, 방벽과 철조망을 설치하여 접근을 통제하고, 경계병이 24시간 출입자를 감시하고 있습니다.
A : 탄약고 관리를 위한 가장 중요한 사항은 무엇입니까?
B : 탄약고에는 탄약 받침대를 설치하여 습기를 방지해야 하며, 통풍이 잘 되도록 통풍구를 설치하여 관리하는 것입니다.

== 관련 표현 ==

- 주기적인 검사 및 치환 작업
- 사계청소, 방화대 설치
- 출입 금지
- 실셈 확인
- 이중 목적 고폭탄

7 Inventory

A : How often do you take inventory
B : We have inventory every other month.
A : What is the purpose of the inventory?
B : It is to check the status of insufficient or excess items every other month.
A : Is the inventory check computerized?
B : A program called FRMS is currently used. It is much faster than processing it by hand.

== Related Expressions ==

- consumption rate
- to turn in surplus items
- logistics
- to record the fluctuation of the assets
- 편성부대자원 관리 프로그램: FRMS(Formation Resource management System)

7 재물조사

A : 얼마나 자주 재물조사를 합니까?
B : 우리는 2개월에 한번 재물조사를 합니다.
A : 재물조사의 목적은 무엇입니까?
B : 격월로 물자의 과부족 현황을 파악하기 위하여 재물 조사를 합니다.
A : 재물조사는 전산화되어 있습니까?
B : FRMS라고 하는 전산 프로그램을 활용하고 있습니다. 수작업에 비하여 작업처리가 훨씬 빠릅니다.

== 관련 표현 ==

- 소모율
- 초과품을 반납하다
- 군수
- 자산의 증감을 기록하다

Unit 9. Combat Development

1 Doctrine

A : We have reviewed the blueprint of the Army. Where do you draw the ideas and necessities from?
B : To do so, we have the combat development suggestion system. It is divided into doctrine, structure and organization, training, leader development, and resource management.
A : What does the doctrine section do?
B : It is designed to revise the current doctrine and provide ideas for the development of future doctrine. Thus, it is necessary to keep good records of ideas that are proposed, at any time, through the command. The good idea providers are rewarded.

== Related Expressions ==

- future battlefield environment
- weapons system
- air-land battle
- information society, scientific thinking

Unit 9. 전투발전

1 교리

A : 장차전에 대비한 육군의 청사진을 알아보았는데, 이를 위한 소요와 아이디어는 어디에서 발굴합니까?

B : 그런 목적으로 전투발전 요구 제안 제도를 시행하고 있습니다. 이는 교리, 구조 · 편성, 무기 · 장비, 교육훈련, 간부계발, 자원관리로 나누어집니다.

A : 교리분야는 어떠한 내용들에 관한 것입니까?

B : 기존 교리의 보완 및 미래 교리의 개발입니다. 따라서, 평소 아이디어의 기록이 필요하고 계통을 통하여 수시로 제안을 하고 있습니다. 좋은 착상들은 적절한 포상도 하고 있습니다.

== 관련 표현 ==

- 미래 전장환경
- 무기 체계
- 공지전
- 정보화 사회, 과학적 사고

2 Structure and Organization

A : What is the direction of the structure and organization in combat development?
B : To the technology-intensive, scientific, and mobile structure.
A : Which troops will be modernized in that direction?
B : mechanized corps, aviation operations command, special operaions command, and mechanized divisions.
A : What units are going to be mechanized?
B : These are the corps, infantry divisions, and special purpose units.
A : Have you made specific efforts to do that?
B : Yes, the TRADOC, military related institutes, and various personnel are concentrating efforts to do that.

== Related Expressions ==

- intelligence function
- reduction in force(RIF)
- in accordance with the national security concept
- force modernization
- Training and Doctrine Command(TRADOC)

2 구조 및 편성

A : 구조 · 편성 분야의 전투발전 방향은 무엇입니까?
B : 기술 집약형의 정보화, 과학과, 기동화 구조로 발전시키는 것입니다.
A : 어떤 부대가 그 방향으로 현대화됩니까?
B : 해당 부대는 기계화 군단, 항작사, 특전사, 기계화 사단 등입니다.
A : 기계화에 중점을 두는 부대는?
B : 일반 군단, 보병 사단, 그리고 특수 목적 부대입니다.
A : 이에 대한 구체적인 노력은 이루어지고 있습니까?
B : 예, 교육사를 비롯한 군 관련 기관 및 다양한 인원들의 노력이 집중되고 있습니다.

== 관련 표현 ==

- 정보 기능
- 병력 감축
- 국가안보 개념에 부합하다
- 군 현대화
- 교육 사령부

3 Weapons and Equipment

A : Modernization of weapon systems is necessary in other to realize the developed concepts.
B : That's correct. This kind of proposal falls on the weapons and equipment section. Modernization of current weapons is essential.
A : Considering the increased significance of intelligence collection, the modernization and future development of all intelligence assets are in demand, aren't they?
B : That's right. There are many areas which should be developed. Detecting the enemy's equipment and improving intelligence circulation devices are two critical areas.

== Related Expressions ==

- surveillance equipment
- imagery intelligence
- strategic intelligence
- collection, analysis, process, dissemination, and usage
- signal intelligence
- unmanned aerial vehicle(UAV)

3 무기 및 장비

A : 발전된 교리 개념을 구현하려면 무기 체계 개선도 필요할텐데요.
B : 예, 이러한 분야에 대한 소요제기는 무기 · 장비에 해당합니다. 기존 무기의 현대화는 필수적입니다.
A : 정보수집의 중요성이 커짐에 따라, 정보무기들의 개선과 발전이 요구됩니다. 그렇죠?
B : 맞습니다. 적을 찾는 장비로부터 정보를 유통시키는 장비 및 유통 체계까지 두 개의 주요한 분야가 있으며 발전의 여지가 많습니다.

== 관련 표현 ==

- 감시 장비
- 영상정보
- 전략정보
- 수집, 분석, 처리, 전파, 그리고 사용
- 신호정보
- 무인 정찰기

4 Training and Education

A : How do you classify the current training of troops?
B : It is classified into unit training, individual training, and reserve force training.
A : What do you emphasize in these types of training?
B : The key points are performing the combined and joint operations with unit training, developing the best soldiers through enlisted training, and improving implementation of combat functions and employing reserve combat power in all reserve force training.
A : What has changed in reserve force training?
B : The major changes are the operating concept, training systems, scientific methods, and introduction of the Total Training Complex.

== Related Expressions ==

- Mater Scenario Event List(MSEL)
- mobilization designee
- military occupational specialty(MOS)
- training circular(TC)

4 교육 훈련

A : 교육훈련은 어떻게 분류하고 있습니까?
B : 부대훈련, 병 훈련, 예비군 훈련으로 분류하고 있습니다.
A : 위의 훈련들은 어디에 중점을 두고 있습니까?
B : 부대훈련은 연합 및 합동작전 수행, 병 훈련은 전투원의 정예화, 예비군훈련은 예비전력의 전투수행능력 향상에 중점을 두고 있습니다.
A : 예비군훈련의 변경 사항은 무엇입니까?
B : 예비군 운용개념, 훈련체계, 과학화 기법, 종합훈련 센터 설치 도입 등이 주요 변경 사항입니다.

== 관련 표현 ==

- 주요사태 목록
- 동원 대상자
- 군사 주특기
- 교육회장

  * 신교리 및 개선교리 등을 포함하는 회람형식의 교리문헌. 교육회장은 신속한 전파가 요구되는 경우에 발간되며 수정을 거쳐 영구적인 교리로 채택되거나 부적합 시에는 폐기됨

5 Leader Development

A : What is the concept of leader development?

B : It is designed to develop the leader's capability to defeat the enemy in the battlefield.

A : What are the expected levels of leader development?

B : It is divided into three levels; specialization and instructor level (mastery), going into action and putting into practice level (awareness), and comprehension level.

A : Who is responsible for the leader development?

B : The primary responsibility goes to the leader himself. At the same time, the commandants and instructors of taining units and the field commanders are also responsible for the leader development of the subordinates.

== Related Expressions ==

- The army should be led by the leaders.
- The success or failure of the army organization depends on the strength or weaknesses of the leader.
- Professional development of the leader is the core project to nurture combat power.
- Outstanding leaders are not made in a day.

5 간부계발

A : 간부계발의 개념은 무엇입니까?
B : 군 조직의 핵심인 간부를 현재 및 미래의 전장에서 적과 싸워 이길 수 있는 능력을 계발해 나가도록 하는 것입니다.
A : 간부계발의 요망 수준은?
B : 요망수준은 전문화 및 교관 수준(숙달), 행동화 및 실천 수준(숙지), 이해 수준의 3 등급으로 구분되어 있습니다.
A : 간부계발의 책임은 누구에게 있습니까?
B : 기본적인 책임은 간부 스스로에게 있습니다. 동시에 교육기관의 장 및 교관과 실무 부대 지휘관도 부하들의 간부계발에 대한 책임을 집니다.

== **관련 표현** ==

- 군은 간부에 의해 주도되어야 한다.
- 군 조직의 승패는 간부 능력의 강약에 따라 좌우된다.
- 간부능력계발은 전투력 육성의 핵심과업이다.
- 우수한 간부는 하루아침에 육성되지 않는다.

6 Resource Management

A : What do you mean by resource management?
B : It means to update the management of personnel, logistics, and finance by using computer systems.
A : In which direction is personnel management moving?
B : It is going to revise the personnel structure into the leader-centered, technology- oriented model. With the integrated defense computer net, we will establish the new personnel management system. In addition, the pre-notification assignment system will be stabilized, and we'll continuously improve the working conditions, morale and welfare systems.
A : Consequently, efficient combat development will be realized through high technology and be the result of a more economic and efficient resource management system.

== Related Expressions ==

- automated resources management system
- integrated logistics analysis program

6 자원관리

A : 자원관리는 무엇을 의미합니까?
B : 그것은 인력, 군수, 예산관리의 정보화를 뜻합니다.
A : 인력분야 발전방향은 무엇입니까?
B : 인력분야는 간부중심 기술집약형으로 인력구조를 개선, 국방전산망과 연계하여 인력관리 시스템을 구축, 사전 보직예고제의 정착, 복무여건, 사기와 복지를 지속적으로 개선하는 것입니다.
A : 결과적으로 과학화된 자원관리체계를 구축하여 경제적이고 효율적으로 자원을 관리하기 위한 전투발전을 뜻하는 것이군요.

== **관련 표현** ==

- 자원 관리 자동화 체계
- 통합 군수 분석계획

Unit 10. War History

1 Nakdong River Battle

A : Why was the Nakdong River Campaign so important?
B : It was the turning point for the ROK Army and the U.N. Forces which were losing at the beginning of the Korean War.
A : What were the major reasons for the ROK Army's victory in the Nakdong River?
B : First, the ROK Army's decisive determination to fight to its death. Second, the U.N. Forces' effective maneuvering ability.

== Related Expressions ==

- to conduct a hasty river-crossing
- carpet bombing
- the salient of the Nakdong River
- to reach the operational limit
- the enemy's massive offensive, three times in a row

Unit 10. 전사

1 낙동강 전투

A : 왜 한국전쟁에서 낙동강 지구 전투가 중요합니까?
B : 한국전쟁 초기에 열세였던 한국군과 유엔군이 공세이전하는 역사적 분기점이었습니다.
A : 낙동강 지구 전투에서 한국군이 승리한 요인은 무엇이었습니까?
B : 첫째, 한국군의 결사항전 정신입니다. 둘째, UN군의 효과적인 기동 능력입니다.

== **관련 표현** ==

- 급속도하를 시도하다
- 융단폭격
- 낙동강 돌출부
- 작전 한계점에 도달하다
- 세 차례에 걸친 적의 대규모 공세

2 Incheon Landing

A : What were the dominant factors for the success of the Incheon landing?

B : Based on the principles of war, one of the factors was surprise. At that time, General MacArthur took the initiative by selecting Incheon, the place where the North Korean Army least expected an attack and he overruled the objections from his staff.

A : Why did the staff object to the landing in Incheon?

B : It was not favorable terrain for a large-scale amphibious operations due to the big difference between the rise and fall of the tide.

== Related Expressions ==

- to apply the principles of war
- to secure a bridgehead
- to strike the enemy at an unexpected time and place
- naval gunfire
- Marine Corps
- to cut the enemy's line of communications(LOC)

2 인천 상륙 작전

A : 인천 상륙 작전의 성공요인은 무엇입니까?
B : 전쟁원칙 측면에서 볼 때 기습입니다. 당시 맥아더 장군은 북한군이 예상치 못하였고, 참모들이 반대한 인천을 상륙지점으로 선정하여 전쟁의 주도권을 장악하였습니다.
A : 왜 참모들이 인천으로 상륙하는 것을 반대하였습니까?
B : 인천은 조수간만의 차가 심해서 대규모 부대가 상륙하기에 적합한 지형이 아니었습니다.

== **관련 표현** ==

- 전쟁원칙을 적용하다
- 교두보를 확보하다
- 적이 예상치 못한 시간과 장소에서 타격하다
- 함포사격
- 해병대
- 적의 병참선을 차단하다

3 Vietnam War

A : Did your unit participate in the Vietnam War?
B : We joined the war as part of the main force. The real combat experience was of great help for the development of the division's combat power.
A : What were the difficulties encountered during the combat?
B : The terrain and climate were totally different from Korea. Especially, we had difficulties in responding to the Vietnamese unconventional warfare.
A : How do you make use of the lessons learned in the Vietnam War?
B : We train the forces as realistically as possible based on the axiom that we fight as we are trained; especially security operations in the rear areas.

== Related Expressions ==

- engage in war
- to take advantage of the terrain
- to set up a concealed position
- mines and booby traps
- cave mopping-up operations
- to disperse defoliants

3 베트남 전쟁

A : 귀 부대는 베트남전에 참전하였습니까?
B : 우리는 주력부대로 참전하였습니다. 실전경험이 우리 사단의 전투력 향상에 큰 도움이 되었습니다.
A : 전투 시 주요 애로사항은 무엇이었습니까?
B : 한국과 전혀 다른 지형과 기상이었습니다. 특히 베트콩의 비정규전 전법에 효과적으로 대응하는 데 어려움이 많았습니다.
A : 월남전의 교훈을 어떻게 적용하고 있습니까?
B : 실전과 같이 훈련하고 훈련한 대로 전투한다는 신념으로 평시 부대훈련 특히 후방지역 경계작전에 최선을 다하고 있습니다.

== **관련 표현** ==

- 전쟁에 개입하다
- 지형적 이점을 활용하다
- 은거지를 구축하다
- 지뢰 및 부비트랩
- 동굴 소탕작전
- 고엽제를 살포하다

4 Gulf War

A : What were Saddam Hussein's miscalculations in the Gulf War?
B : First, he underestimated the allied forces' combat capability. Second, he didn't fully recognize that the Iraqi defense positions in Kuwait were exposed. Third, he completely underestimated the allied forces' state-of-the-art weapons.
A : What were the factors in the allied forces' victory?
B : First, it conducted well-coordinated combined operations to defeat the Iraqi forces. Second, it had sufficient resources to support the ground attack and maintained effective combat service support. Third, it had enough time to achieve the highest level of combat power.

== Related Expressions ==

- to estimate the situation prudently
- to overestimate its own combat power
- to train well with conventional weapons and equipment
- The ground component forces will conduct the attack.
- to conduct military operations in foreign countries
- to be equipped with precision guided weapons and computerized fire control equipment

4 걸프전

A : 걸프전 시 후세인의 판단 착오는 무엇이었습니까?
B : 첫째, 연합군의 전력을 과소평가했습니다. 둘째, 쿠웨이트 전역 중 방어망 자체가 이라크 사막에 노출되어 있다는 점을 무시했습니다. 셋째, 후세인은 첨단무기를 완전히 과소평가했습니다.
A : 다국적군이 승리한 요인은 무엇입니까?
B : 첫째, 다국적군은 이라크를 타도하기 위하여 잘 조화된 연합작전을 수행하였습니다. 둘째, 지상공격을 뒷받침할 수 있는 풍부한 자원을 가지고 있었고, 효율적인 전투근무지원을 유지하였습니다. 셋째, 최고의 전투력 수준을 유지하기 위한 시간적 여유가 충분하였습니다.

== **관련 표현** ==

- 신중하게 상황판단을 하다
- 자신의 전투력을 과대 평가하다
- 재래식 무기와 장비를 잘 훈련시키다
- 지상구성군 부대가 공격을 감행하다
- 외국에서 군사작전을 실시하다
- 정밀 유도무기와 컴퓨터화된 사격통제장비를 구비하다

5 Yeonpyeong Naval Engagement

A : I heard that there was a recent sea skirmish with the North Korean Navy.

B : We defeated the North Korean naval patrol boats in the Yellow Sea near Yeonpyeong Island.

A : What was the actual capability of the North Korean Navy?

B : They were not as strong as we thought. Therefore we could destroy them without any strong resistance.

A : Did you predict the North Korean's provocation?

B : We knew the features and locations of the North Korean vessels. The Naval Fleet Commander sent real-time information. The intelligence was promptly forwarded to the Joint Chiefs of Staff.

== Related Expressions ==

- Northern Limit Line(NLL)
- fleet
- destroyer
- submarine
- naval gunfire

5 연평해전

A : 최근에 북한과 교전하여 승리한 전투가 있다고 들었습니다.
B : 서해 연평도에서 침투한 북한의 전대를 격파하였습니다.
A : 실제로 북한의 해군전력은 어떤 수준이었습니까?
B : 우리가 예상했던 것보다는 강력하지 않았습니다. 그래서 우리 해군은 큰 저항 없이 격파할 수 있었습니다.
A : 사전에 북한이 도발하리라고 예상하였습니까?
B : 한국군은 적 함정의 특성과 제원에 대하여 잘 파악하고 있었습니다. 함대 지휘관은 실시간 첩보를 보내왔으며, 그 정보는 합참에 신속하게 제공되었습니다.

== **관련 표현** ==

- 북방 한계선
- 함대
- 구축함
- 잠수함
- 함포

Unit 11. Learn English with Pop Songs

1 I have a dream : 나에겐 꿈이 있어요

Sung by ABBA

I have a dream, a song to sing
내겐 꿈이 있고, 부를 노래도 있어요.

To help me cope with anything
어떤 어려움도 이겨낼 수 있게 도와주지요

If you see the wonder of a fairy tale
만약 동화 같은 이야기를 믿는다면

You can take the future even if you fail
아무리 힘들어도 미래를 꿈 꿀 수 있어요

I believe in angels
난 천사를 믿어요.

Something good in everything I see
내가 보는 모든 사물들의 좋은 점을 보며

I believe in angels when I know the time is right for me

난 천사를 믿어요 적당한 때가 오면

I'll cross the stream - I have a dream
저 강을 건널 거라는 걸 난 알아요 난 꿈이 있거든요
I have a dream, a fantasy
난 꿈이 있어요 환상이 있어요
To help me through reality
현실을 헤쳐나가게 해주고

And my destination makes it worth the while
내 운명이 그것을 가치 있게 만들어 줘요

Pushing through the darkness still another mile
아직도 남아 있는 어두움을 헤쳐 나갈 수 있게 해줘요

I believe in angels
난 천사를 믿어요.

Something good in everything I see
내가 보는 모든 사물들의 좋은 점을 보며

I believe in angels
난 천사를 믿어요

When I know the time is right for me
적당한 때가 오면

I'll cross the stream - I have a dream
저 강을 건널 거라는 걸 난 알아요 난 꿈이 있거든요

I'll cross the stream - I have a dream

저 강을 건널 거라는 걸 난 알아요 난 꿈이 있거든요

I have a dream, a song to sing
내겐 꿈이 있고, 부를 노래도 있어요

To help me cope with anything
어떤 어려움도 이겨낼 수 있게 도와주지요

If you see the wonder of a fairy tale
만약 동화같은 이야기를 믿는다면

You can take the future even if you fail
아무리 힘들어도 미래를 꿈 꿀 수 있어요

I believe in angels
난 천사를 믿어요.

Something good in everything I see
내가 보는 모든 사물들의 좋은 점을 보며

I believe in angels
난 천사를 믿어요

When I know the time is right for me
적당한 때가 오면

I'll cross the stream - I have a dream
저 강을 건널 거라는 걸 난 알아요 난 꿈이 있거든요

I'll cross the stream - I have a dream
저 강을 건널 거라는 걸 난 알아요 난 꿈이 있거든요

2 I owe you love : 너한테 빚을 졌어

sung by Carry and Ron

You believe that I've changed your life forever
당신은 믿고 있어요, 내가 당신의 삶을 영원히 바꾸어 놓았다고

And you're never gonna find another somebody like me
또한 나 같은 사람은 다신 찾지 못 할 것이라고 말이죠

And you wish you had more than just a lifetime
그리고 당신은 당신의 수명보다 더 오래 살기를 바라지요.

to give back all I've given you
제가 당신에게 베푼 것을 돌려주기 위해서

And that's what you believe
그런데 그것은 당신이 그렇게 믿을 뿐입니다.

But I owe you
하지만 저는 당신께 빚지고 있어요

the sun light in the morning
아침의 햇살과

and the nights of all this loving
이 모든 사랑스런 밤들

that time can't take away
시간이 빼앗아 갈 수 없는

And I owe you
저는 당신께 빚지고 있어요

more than life now, more than ever
현재의 삶 이상의 것을, 그 어느 때 보다도 지금

I know that it's the sweetest debt
그것이 가장 달콤한 빚이라는 것을 알아요

I'll ever have to pay
내가 영원히 갚아야 할

I'm amazed when you say
당신이 말할 때 저는 놀랐어요

it's me you live for
당신이 바로 날 위해 산다고

You know that when I'm holding you
당신은 아시죠, 제가 당신을 포옹하고 있으면

you're right where you belong
당신은 바로 있어야 할 곳에 있다는 것을

And my love, I can't help but smile
당신, 저는 미소를 보내지 않을 수 없습니다.

with wonder when you tell me
놀라움으로, 당신이 내게 말할 때

all I've done for you

제가 당신을 위해 한 모든 것을

Cause I've known all alone
왜냐하면 나는 지금까지 혼자만 알고 있었기에

3 Yesterday : 예전엔

sung by Beatles

Yesterday, all my troubles seemed so far away
예전엔, 나의 모든 시름들이 멀리 사라져 버린 듯 했는데

Now it looks as though they're here to stay
이제 그 시름들이 마치 여기 내 곁에 다시 밀려오는 것 같아요

Oh, I believe in yesterday
아, 그때가 좋았었는데

Suddenly, I'm not half the man I used to be
갑자기, 예전 내 모습의 반에도 미치지 못하는 사람이 된 것 같아요

There's a shadow hanging over me
어두운 그림자가 내게 드리워지고 있어요

Oh, yesterday came suddenly
아, 갑자기 지난 날의 추억들이 밀려와요

Why she had to go I don't know
왜 그녀가 떠나야 했는지 난 몰라요

she wouldn't say

그녀는 아무 말도 하려하지 않았어요

I said something wrong
나는 뭔가 잘못된 거라고 말했죠

Now I long for yesterday
지금 이 순간 지난 날이 자꾸만 그리워져요

Yesterday, love was such an easy game to play
예전엔, 사랑은 아주 쉬운 게임 같았어요

Now I need a place to hide away
이제 난 어디든 숨을 곳이 필요해요

Oh, I believe in yesterday
아, 그때가 좋았었는데

Why she had to go I don't know
왜 그녀가 떠나야 했는지 난 몰라요

she wouldn't say
그녀는 아무 말도 하려하지 않았어요

I said something wrong
나는 뭔가 잘못된 거라고 말했죠

Now I long for yesterday
지금 이 순간 지난 날이 자꾸만 그리워져요

Yesterday, love was such an easy game to play
예전엔, 사랑은 아주 쉬운 게임 같았어요

Now I need a place to hide away
이제 난 어디든 숨을 곳이 필요해요

Oh, I believe in yesterday
아, 그때가 좋았었는데

Mm mm mm mm mm mm mm
음음음음 음 음음

PART Ⅱ. Military Briefings

Unit 1. Self-Introduction
Unit 2. NK's Threat and Combat Strength
Unit 3. The Infantry Division
Unit 4. The Fleet
Unit 5. The Flight Wing
Unit 6. Panmunjom
Unit 7. Dora Observation Post
Unit 8. Interview Samples
Unit 9. General Affairs
UNIT 10. Progression of Meeting
UNIT 11. UN Introduction

Unit 1. Self-Introduction

자기 소개

Key Words

commissioned	임관하다	Korea National Defense University	국방대학교
Lieutenant Colonel	중령		
Joint Chiefs of Staff	합동참모본부	security policy	안보정책학
second lieutenant	소위	Airborne Special Warfare Brigade	공수특전여단
company commander	중대장		
operational officer	작전장교	Infantry Division	보병사단

Text

I would like to introduce myself.
제 소개를 드리겠습니다.

I am Lieutenant Colonel Jeon Wan-young of the Overseas Deployment Division of the Operations Headquarters of the Joint Chiefs of Staff.
저는 합동참모본부 작전본부 해외파병과 전완영 중령입니다.

I graduated from 00 University in 1988 and was commissioned as an infantry second lieutenant.
저는 1988년도에 00대학교를 졸업하고 보병소위로 임관하였습니다.

In 1993, I was assigned to the 122nd Division as an infantry company commander.

1993년 제122사단에서 보병 중대장으로 보직되었습니다.

Then served as a 3rd company commander in the 156th Infantry Regiment.

그리고나서 보병156연대에서 3중대장직을 수행했습니다.

After completing the company commander, I served as an operational officer in the 156th Regiment of the 122nd Infantry Division, and from 2003 to the end of 2005, I studied security policy at the Graduate School of Korea National Defense University and obtained a master's degree.

중대장을 마친 다음 제122보병사단 156연대에서 작전장교를 했었고, 2003년도부터 2005년 말까지 국방대학교 대학원에서 안보정책학 공부하여 석사학위를 취득했습니다.

After graduating from the Army College in 2001, he worked as a staff member for the 21st Airborne Special Warfare Brigade.

2001년 육군대학 수료 이후 제21공수특전여단에서 인사참모직을 수행하였습니다.

The battalion commander was in the 137th Infantry Division, and after completing the battalion commander, he served as the operational staff.

대대장은 제137보병사단에서 하였으며 대대장을 마친 후, 작전참모직을 수행했습니다.

I transferred to the Joint Chiefs of Staff here in December 2008 and am working hard with pride that I am working in the highest war command unit.

이곳 합동참모본부에는 2008년 12월에 전입 와서 최고 전쟁지휘 부대에서 근무한다는 자부심을 가지고 열심히 근무하고 있습니다.

I have a lovely wife and two sons. My sons are the 16th and 17th grade.
가족은 고등학교 1학년과 2학년에 다니는 사랑스런 두 아들과 아내가 있습니다.

My family and I always try to spend time together during the weekends, going to the theater, movie, or going on a picnic near Seoul.
제 가족은 휴일을 항상 함께 보내려고 노력하기 때문에 주말에는 연극 혹은 좋은 영화 관람 또는 서울 근교 소풍 등으로 지내고 있습니다.

I also enjoy hiking and my favorite sport is soccer.
저는 또한 등산을 즐기며 가장 좋아하는 운동은 축구입니다.

Unit 2. NK's Threat and Combat Strength

북한의 위협과 군사력

Key Words

stake	위태로운	torpedo boat	어뢰정
aggravate	악화시키다	mine sweeper	소해정
blossom	꽃, 번성하다	theater ballistic missile	전구탄도미사일
devastating	파괴적인,	chemical decontamination unit	화생방 제독부대
nuclear weapon	핵무기	personnel strength	병력
freeze	동결, 금지	vulnerability	취약점
agreement	협정	belligerent	적대적인, 호전적인
Negotiation	협상	deterrence	억제
dwindling	줄어드는		

Briefing Text[4)]

Good morning, I consider if an honor to have the opportunity to brief you on nK's threat and combat strength. This is a four part briefing that with highlights of the geo-political and economic significance of the region.

4) 연합사 『Practical Military English Ⅱ』(비매품), 『국방백서 2014』를 참조하여 재구성한 것임.

안녕하십니까? 여러분을 모시고 북한의 위협과 군사력에 관한 브리핑 기회를 갖게 된 것을 기쁘게 생각합니다. 본 브리핑은 4개의 분야로 구성되어 있으며 이 지역이 갖는 지정학적 및 경제적 의의로부터 시작되겠습니다.

First of all, Northeast Asia is a complex and dynamic region. It is important for people working in the region to appreciate the historic relationships that exist between the regional actors. The stability of the region and the security of vital national interests are at stake. Taiwan has a strong economy, yet recently aggravated tensions between an and the People's Republic of China(PRC) have come to the world's attention. China is solidifying its economic base and is seeking spread its influence into the void left by the former Soviet Union. Russia possesses vast supplies of natural resources but is struggling with democracy and trying to survive economically.

먼저 동북아시아는 복잡하고 역동적인 지역으로 이곳 사람들에게는 국가 간의 역사적인 관계가 중요하게 작용하고 있습니다. 이 지역의 안정과 국가들의 안보는 지금 위험에 처해 있습니다. 대만의 경제력은 강하지만, 최근 대만과 중국 사이의 긴장은 급속히 악화되어 세계의 이목을 집중시키고 있습니다. 중국은 경제적 기반을 총집결시켜 그의 영향력을 구소련의 공백으로 확산시키려 하고 있습니다. 러시아는 막대한 천연자원을 보유하고 있지만 민주주의와의 투쟁과 함께 경제적으로 생존하기 위한 힘든 노력을 계속하고 있습니다.

Like Taiwan, Japan is an economic power not only in Asia but throughout the world. And, in the center of Northeast Asia is Korea, The Republic of Korea has a blossoming democracy and the country a key trade partner in the region. The security of the Republic of Korea is of vital interest to the United States and the United Nations.

대만과 마찬가지로 일본도 아시아 내에서 뿐만이 아니라 세계적으로도 강한 경제력를 가지고 있는 나라입니다. 그리고 동북아시아의 중심에 대한민국이 있습니다. 대한민국은 민주주의의 발전을 이루었으며, 이 지역의 주요 무역 국가로 떠올랐습니다. 한국의 안보는 미국과 유엔에 아주 중요한 요소가 되었습니다.

Other than all these economic matters, an important fact related to regional stability is that four of the five largest militaries in the world are represented within one thousand kilometers of Seoul. The Chinese, [he Russians, the U.S. presence in South Korea, and the fifth largest military is in nK. nK's force is capable of launching a devastating Offensive with limited warning.

경제 외에도 지역 안정에 관련해서 중요한 것은 세계의 5대 군사력 중 4개가 서울에서 1,000km 이내에 위치되어 있다는 것입니다. 중국, 러시아 주한미군, 그리고 5번째 군사력을 가지고 있는 북한입니다. 북한은 제한된 경고 하에 파괴 공격을 할 수 있는 능력을 가지고 있습니다.

Once again, regional stability is interdependent with economic stability, political stability, and military stability of all the factors in the region.

History shows that if a militarily capable regime lacks economic resources, then there is political pressure on the regime to use their military to acquire needed resources. Potentially, nK could follow this historical model.

다시 한 번 말씀드리자면, 지역의 안정은 지역 내 모든 국가들의 경제적, 정치의 그리고 군사적 안정과 상호 의존적이라고 할 수 있습니다. 만약 군사력을 보유 정권이 경제적 자원이 부족하다면, 필요한 자원을 획득하기 위해 군사력을 사용 하라는 정치적 압력이 정권에 가해진다는 것을 역사는 보여 주고 있습니다. 그러므로 북한이 이러한 역사적 패턴을 따라갈 수 있다는 것을 배제할 수 없습니다.

The DPRK's of nuclear weapons development program is a major great to the stability of the region and is currently the focus of political tension on the peninsula. In 1994, the US and nK entered |to an agreement called the AGREED FRAMEWORK. Under this agreement, nK agreed to freeze its nuclear program, allow International Atomic Energy Agency(IAEA) inspections, and to comply with 1992 north-South Denuclearization Agreement. In exchange, nK will receive light water reactor(LWR) facilities and economic cooperation, mainly in the form of fuel oil.

북한의 핵무기 개발계획은 이 지역의 안정에 큰 위협이며 현재 한반도에서의 정치적 긴장의 초점이 되고 있습니다. 1994년에 미국과 북한은 "기본 원칙 동의" 관한 협정을 체결했습니다. 이 협정에 의하면 북한은 핵 개발을 중지하기로 했고 국제원자력기구의 사찰과" 1992년에 이루어진 남북한 핵 개발 중지 협정에 따르기로 했습니다. 그 대가로 북한은 두 개의 경수로와 연료를 주로 한 경제적인 지원을 받기로 했습니다.

An important part of this agreement was the formation of the international consortium called the Korean Peninsula Energy Development Organization or KEDO. KEDO provides the financing, construction and technology for the light water reactors.

이 협정의 중점은 한반도 에너지개발기구의 설립이었습니다. 한반도 에너지개발기구는 경수로에 대해 자금과 건설, 그리고 기술을 제공합니다.

However, this process is behind schedule because of suspected nK non-compliance with the agreement. Negotiations are currently underway for US inspectors to gain access to the Kum-chang-ni site, a suspected nuclear facility. So by May 1st 1999, the Executive Branch of the US Government must certify to Congress that nK has indeed frozen their nuclear program. If nK is certified as compliant, then hopefully, Congress and KEDO members(US, ROK, Japan) will be more timely with appropriations for continued work on the LWR and economic cooperation.

그러나 이 과정은 북한의 협정 불이행 의혹으로 인해 지연되고 있습니다. 현재 미국과 북한은 핵 의혹 시설인 금창리에 대한 사찰 협상 중에 있습니다. 1999년 5월 1일까지 미 정부는 미의회에 북한이 핵 계획을 동결했다는 것을 증명해 보여야 합니다. 만약 북한이 핵 협정을 이행했다고 증명될 경우에는 미의회와 한반도 에너지개발기구 회원국들(대한민국, 미국, 일본)은 다시 잔여분의 경수로와 경제적 지원을 북한에 제공할 계획입니다

Even though there is an effort to help nk, political and economical tensions still exist on the Korean peninsula. The republic of Korea's expanding economy reflected a growth rate exceeding not only European nations, but japan as well. In contrast, nK's economy took a negative turn in 1990. However, the greatest burden on nK's economy is self imposed because they still spend 26% of their dwindling GDP on building the nK military.

북한을 돕자는 많은 노력에도 불구하고 지금까지도 한반도에는 정치적, 경제적 긴장이 존재하고 있습니다. 대한민국의 경제발전은 성장률에서 유럽뿐만 아니라 일본까지도 앞질렀지만, 그 반대로 북한의 경제는 1990년을 기점으로 마이너스 성장을 시작했습니다. 그러나 북한의 가장 큰 경제적 문제점은 북한 스스로에게 있습니다. 왜냐하면 북한은 계속 그들의 감소하고 있는 국내 총생산의 26% 군사력에 투자하기 때문입니다.

Next, I will discuss nK's combat strength. to begin with, there are 600.000 nK soldiers located south of pyongyang-wonsan line, and its ground troops are estimated at 1,000,000 personnel on active duty and 6,000,000 reserves. other nK's army weapon assets are 10,000 tanks 2,000 APCs 2,500 MRLs 10,000 artillery pieces and 12000 ADA guns.

다음은, 북한의 전투력에 대해서 알아보겠습니다. 먼저 약 600,000명의 북한군들이 평양-원산선의 남쪽에 배치되어 있으며, 또, 약 1,000,000명의 현역과 6,000,000명의 예비군을 보유하고 있습니다. 그리고 북한이 보유하고 있는 육군의 무기들을 살펴보면, 10,000대의 탱크, 2,000대의 장갑차, 2,500대의 다련장 로켓포, 10,000대의 포병화기, 그리고 12,000대의 방공포병화기 등이 있습니다.

Of these troops, about 80,000 are Special Operation Forces, whose mission is to recon, to support conventional operations, to establish a second front in enemy's rear area to, counter enemy SOF in nK and to maintain internal security.

그 중에서 약 80,000명 규모의 특수작전부대가 있습니다. 이들의 임무는 수색 재래식 작전지원, 적 후방에 제2전선 형성, 북한 내에서의 특수작전부대에 대응 그리고 내부 안보 유지 등이 되겠습니다.

Next is the combat strength of nK's navy and Air force. First the location and strength of nKs Naval Organizaion Fleet Command Headquarters are located at Toejo Dong and Nampo, and Naval Bases are located at Najin, Wansan and Pipagot. There are 50,000 personnel 1,000 plus vessels torpedo boats, coastal patrol patrol boats, and mine sweepers. These are all a threat to sea lines of communication.

다음은 북한의 해군과 공군의 전투력입니다. 먼저 해군기지의 위치와 전투력입니다. 북한의 함대사령부는 퇴조동과 남포에 위치하고 있고, 해군기지들은 남진, 완산, 비파곶에 위치되어 있습니다. 그리고 북한 해군 병력은 약 50,000명이며 1,000척(잠수함, 상륙함, 어뢰정, 해안경비정, 해함) 이상의 배들을 소유하고 있습니다. 그리고 이들은 해상 병참선에 위협을 줄 것입니다.

Following is information regarding nK's Air Force. There are about 100,000 personnel and 1,500 aircraft, including 800 tactical fighters(MIGs), 100 bombers, 300 transports, and 300 helicopters in the nKAF inventory.

계속해서 북한 공군에 대해 보고드리겠습니다. 북한의 공군 병력은 100,000명이며, 1,500대의 비행기(800대의 전투기, 100대의 폭격기, 300대의 수송기, 300대의 헬리콥터)를 보유하고 있습니다.

Other than nK's Army, Navy, and Air Force, there are theater ballistic missiles, chemical weapons, and nuclear capabilities which would leave a great number of casualties, if a war broke out. There are three kinds of theater ballistic missiles. SCUD missiles have a range of 500Km, No Dong missiles which can reach as far as 1,000 Km, and lastly Dae Po Dong missiles which can strike a target at 1500-plus kilometers. Also, nK has chemical capabilities, such as production capability, chemical decontamination units, and weapon Systems capable of delivering chemical munitions. And finally, nK has Soviet supplied research reactors such as 10-30MW reactors, 50-200MW reactors, and reprocessing plants.

북한의 육 · 해 · 공군 이외에도 전구 탄도미사일, 화학무기 그리고 핵 능력 등은 전쟁 발발 시 대규모 사상자를 유발시킬 수 있습니다. 북한은 세 가지 유형의 전구탄도미사일을 보유하고 있습니다. 500Km 사정거리의 스커드 미사일과 1,000Km · 사정거리의 노동미사일 그리고 마지막으로 1,500Km 이상의 사정거리를 가지고 있는 대포동미사일이 있습니다. 북한은 이외에도 화학무기의 생산, 제독부대 그리고 화학무기 탄두를 발사할 수 있는 무기체계를 가지고 있습니다. 마지막으로 북한은 소련이 제공한 10-30MW 및 50-200MW급의 연구용 원자로, 그리고 핵 연료 재처리 공장이 있습니다.

If we compare nK's pas decade combat strength to today's there have been lots of changes since the. nKs personnel strength has increased from 6000,000 to 1,100,000 of which 65% have been forwardly deployed also there has been growth in the artillery corps and their combat capability has been modernized.

만약 10년 전의 북한과 오늘날의 북한을 비교해 보면 많은 변화들이 있는 것을 볼 수 있습니다. 북한은 병력을 600,000명에서 1,100,000명으로 증가시켰고 이 중 65% 군사력을 전방에 배치시켰습니다. 또한 포병의 발전이 있었으며 전투력 등이 현대화되었습니다.

Now, if we consider all the factors discussed today's, Seoul, with a population of 13 million, 1/4 of the entire South Korean population, has some vulnerabilities. Being a Historical and cultural COG, it is only 22 miles from the DMZ, it's within range of artillery close, and is vulnerable to sea infiltration. This is why we have to keep our state of Readiness high at all times with the help of US Forces in Korea.

그렇다면 지금까지 말씀드린 내용을 고려해 볼 때, 대한민국 인구의 1/4인 1천3천백만 명이 살고 있는 서울은 몇 가지 취약점들을 가지고 있습니다. 서울은 역사적, 문화적 중심으로 비무장지대에서 약 22마일 거리에 있으며, 적 포 사정거리 안에 있고, 또한 해안침투 등에 취약합니다. 이런 점들이 바로 우리가 주한미군과 협력해서 항상 대비태세를 유지해야 하는 이유인 것입니다.

As I have mentioned before, the Republic of Korea is facing nK, the most belligerent country in the world. While we face nK, CFC's duty is to deter external aggression, and if deterrence fails, to defend the Republic of Korea along with the UNC and to defeat the aggressors, CFC, being the defensive core on the Korean Peninsula, is well-prepared so as to not allow under any circumstances, any external attack so or infiltration of any kind.

이상 말씀드린 바와 같이 대한민국은 지구상에서 가장 호전적인 집단인 북한과 대치하고 있습니다. 이러한 대치 상황에서 한미연합군사령부의 임무는 외부의 적대행위 억제 그리고 억제 실패 시 유엔군사령부와 함께 대한민국을 방어하고, 침략자들과 싸워 이기는 것입니다. 연합사는 한반도 방위의 주축으로서 외부의. 침략 또는 침투 등 모든 가능한 적대행위를 절대 용납하지 않을 것이며 이에 대한 만반의 준비를 하고 있습니다.

I will brief on specific details about CFC's combat power and readiness posture after a 10 minute break.

한미연합군사령부의 전력과 대비태세에 대한 자세한 내용은 약 10분간의 휴식 후 계속해서 보고 드리겠습니다.

Unit 3. The Infantry Division

보병사단 일반현황

Key Words

be activated in	~ 창설되다	Depth	종심
surprise attack	기습공격	Military Demarcation Line.	군사분계선
delay operation	지연작전		
front line	전선, 전방	avenue of approach	접근로
infiltration operation	대침투작전	key terrain	주요지형
infantry regiment	보병연대	infantry battalion	보병대대
aviation troop	항공부대	reserve	예비
Front(frontage)	정면	reinforcement	증원

Briefing Text

Good morning, General Lee, I'm Lieutenant Colonel Kim, the G3, Operations. On behalf of the commanding general and soldiers of the Division, I sincerely welcome you to the 77th Infantry Division, which has 70 years of brilliant history and tradition.

좋은 아침입니다, 이 장군님, 저는 G3 김대령입니다. 사단장님과 장병을 대신하여 70년의 역사와 전통에 빛나는 저희 77사단을 방문해 주신 것을 진심으로 환영합니다.

First, let me tell you our Division history briefly. The Division was activated in Chuncheon in 1948. When the enemy launched a surprise attack in June 1950, the Division successfully conducted delaying operations under the extremely adverse conditions.

먼저 사단의 역사에 대해 설명드리겠습니다. 사단은 1948년 춘천에서 창설되었습니다. 1950년 6월 적의 기습공격 시 불리한 여건에서도 지연작전을 성공적으로 수행했습니다.

Throughout the Korean War, our Division distinguished itself many times. In October 1950, our Division was the first which advanced into Pyongyang, the capital of North Korea.

한국전쟁을 통하여 사단은 여러 차례의 명성을 떨친바 있습니다. 1950년 10월에는 북괴의 수도인 평양에 최초로 입성한 사단이 되었습니다.

During the war, the Division successfully carried out the missions of various offensive and defensive operations in the central part of the front line.

전쟁이 진행되고 있는 동안 사단은 전선의 중앙에서 다양한 공격 및 방어 작전의 임무를 성공적으로 수행했습니다.

On the 10th of April 1952, the Division moved to its present area, where fierce fighting took place during the Korean War. Major General Lee assumed command of the Division in January 2023, as the 52th commander of the Division.

1952년 4월 10일, 사단은 한국전쟁의 격전지였던 현 위치로 이동하였습니다. 현 사단장님이신 이 장군님은 2023년 1월부터 52대 사단장으로 부대를 지휘하고 계십니다.

The missions of the Division, as shown on the screen, are to defend its assigned area of responsibility, to prepare operation plans, to control its assigned DMZ area, to conduct counter infiltration operations, to train and educate members of the unit, and to be prepared to conduct other missions.

차트에서 보시는 바와 같이, 사단의 임무는 지정된 책임지역을 방어하며, 작전계획 준비 비무장지대의 책임 지역 통제, 대침투작전 수행, 부대원의 훈련과 교육, 기타 부여된 임무를 수행하고 있습니다.

The Division consists of three infantry regiments, one artillery regiment with four self-propelled artillery battalions, and troops under direct control of Division: Tank, Reconnaissance, Engineer, Signal. One aviation troop and one TOW company are attached to the Division.

사단은 3개 보병연대, 4개 자주포 대대를 가진 1개의 포병 연대 및 전차, 정찰, 공병, 통신부대로 구성되어 있습니다. 항공부대와 토우중대는 사단에 배속되어 있습니다.

Next, I would like to brief you on the Division's area of responsibility. The Division defends 20km in the central part of the Military Demarcation Line.

다음으로, 사단의 책임 지역에 대해 보고드리겠습니다. 사단은 군사분계선 중앙부에 20km를 방어하고 있습니다.

The corridor in the Division's area of responsibility is one of the most important enemy avenues of approach from North Korea toward the center of gravity of Korea. It is only 70km from here to Seoul and so our Division s practically guarding the gateway to Seoul. Maseok City with a population of 50,000 people is located in the Division's area of responsibility. Major reads are Road No. 3 and Road No. 5. In the Division sector, the Imjin River, with a total length of 10km, flows from east to west.

사단 책임지역은 지형은 북한에서 남한의 중심부에 이르는 가장 중요한 접근로중의 하나입니다. 이곳에서 서울까지는 70km에 불과하므로 사단은 실질적으로 서울의 관문을 수호하고 있습니다. 사단 지역 내에는 인구 50,000명의 마석시가 위치하고 있습니다. 주요 도로로는 3번, 5번 도로가 있습니다. 사단 책임지역 내에는 10km의 임진강이 동쪽에서 서쪽으로 흐르고 있습니다.

Key terrains in our Division's area of responsibility are Hill 550, and Hill 645. The three main enemy avenues of approach into our Division sector are from A to Gosung City, from B to Hill 511, and from C to the eastern flank of the Division's rear area. All of these avenues of approach are favorable to mechanized forces' movement.

사단 지역내의 주요 지형지물로는 550, 645고지가 위치하고 있습니다. 사단지역의 내 3개 주요 적 접근로는 A에서 고성시에 이르는 접근로와, B에서 511고지에 이르는 접근로, C에서 사단 후방 동쪽 측위에 이르는 접근로가 있습니다. 모든 접근로는 기갑부대의 기동이 가능합니다.

The Division opposes the enemy's 5th and 6th Divisions of the 1st Corps. The composition of the enemy Corps appear to be 15 infantry battalions and three tank battalions supported by 120mm artillery and division artillery battalions. Their estimated reinforcements are 3 regiments in reserve 2nd echelon Reserve Regiment, 1 artillery regiment, 2 fllow-on divisions, and 3 armored battalions.

사단은 적 1군단 예하 5사단, 6사단과 대치하고 있습니다. 적의 구성은 120mm 포병과 사단 포병대대의 지원을 받는 15개의 보병대대, 3개의 기갑대대이며, 예상 증원 부대로는 제2제대의 3개 예비연대, 1개의 포병연대, 후속 2개 사단 및 3개의 기갑대대로 판단됩니다.

Currently, the Division has deployed 2 battalions in the GOP area and 4 battalions along the FEBA. 3 regiments from the 2nd Infantry Division are attached to our division as a Division Reserve. When war breaks out, GOP operations can be intensified by turning it into a forward area defense operations.

현재 사단은 GOP에 2개 대대, 그리고 전투지역전단을 따라 4개 대대가 전개되어있으며, 2사단의 3개 연대가 작전통제 되어 사단 예비로 배속되어 있습니다. 전시에 GOP작전은 전방지역 방어작전으로 전환함으로써 보다 강화될 것입니다.

The Division has secured all the key terrains in the GOP area. In order for successful and decisive defensive operations, screening and delaying operations in the western are at of the GOP are essential. By conducting those operations, we would quickly neutralize the enemy's 1st echelon offensive forces and the 2 echelon forces that were committed early.

사단은 GOP 내의 주요 지형을 확보하고 있습니다. 성공적이며 결정적인 방어작전을 위해 GOP 서쪽지역에서의 차장 및 지연작전은 아주 중요합니다. 이러한 작전을 실시함으로써 적의 제1공격제대 및 조기 투입된 제 2제대를 신속히 무력화 시킬 수 있습니다.

The Division has established four kill zones along the enemy's main avenues of approach. The Division has 3 artillery battalions for immediate useg and 5 artillery battalions for additional reinforcement, which enough artillery fire concentration on these kill zones. If the Corps reinforces our firepower with its available artillery, we will have a total of 300 artillery pieces from 8 artillery battalions.

사단은 적의 주접근로를 따라 4개의 살상지대를 설치했습니다. 사단은 즉시 사용할 수 있는 3개 대대와 추가적인 증원을 위한 5개 대대로 포병사격을 집중시킬 수 있습니다. 만약 군단의 가용 포병화력으로 증원이 되면 총 8개 포병대대 300개 포가 되겠습니다.

To hinder enemy tanks, the Division emplaced anti-tank barriers reaching 5km in length and a total of 40 road blocks. The Division constructed 15km of barbed wire entanglements including the interior barrier.

적 전차를 거부하기 위해 사단은 총 5km의 대전차 장애물과 40개의 도로 장애물을 설치했으며 내부 장벽을 포함하여 15km의 윤형 철조망을 설치하고 있습니다.

Personnel strength is as follows: 92% of officers and 95% of enlisted men are assigned. So there shouldn't be any problem to accomplish our assigned mission with this personnel strength. As shown on this chart, tanks and major weapons on hand are at 90%. However, the Division has a shortage of 17 vehicles. This provides us 80% mobility of equipment. However, this shortage of equipment is not critical to the Division's operations. mission

인력현황은 다음과 같습니다. 장교92%, 사병 95%가 보직되어 있으므로 우리에게 부여된 임무를 완수하는데 전혀 문제가 없습니다. 차트에서 보시는 바와 같이 보유하고 있는 전차 및 주요화기는 90% 수준을 유지하고 있으나, 차량은 17대가 부족하여 장비의 기동력은 80% 수준을 유지하고 있습니다.

As briefed above, the Division successfully conducts its mission of providing security and denying enemy infiltration. The Division is also doing its best to conduct realistic military training and to achieve total material readiness. This allows the Division to meet its goal of "The Combat Readiness Posture to knock out the enemy in the first round." All officers and soldiers of the Division will crush the nk Communists during the initial stage of battle. This concludes my briefing. Do you have any questions? Thank you.

이상 보고드린 바와 같이, 사단은 경계 및 적 침투 거부 임무를 수행하고 있습니다. 사단은 또한 실질적인 군사훈련과 모든 물자준비태세를 달성하기 위해 최선을 다하고 있습니다. 따라서 사단은 초전박살 전투태세 확립의 목표를 충분히 달성하고 있습니다. 사단의 전 장병은 초전에 북괴 공산주의자들을 분쇄시킬 것입니다. 이상으로 보고를 마치겠습니다. 질문 있으면 해주십시오. 감사합니다.

Reference

한국식 영작은 이렇게 바꾸자[5)]

한국인이 만든 영어 문장은 크게 세 가지 구조적인 특진을 보입니다. **첫째는 힘들여 명사형을 만들어 상황을 표현하는 경우입니다**. 이런 문장은 아무래도 좀 복잡하고 어색하지요. 이때는 명사로 표현한 내용을 풀어서 동사로 표현하는 편이 낫습니다. 동사가 분명하게 드러나지 않고 행위주가 주어로 쓰이지 않았기 때문에 문법적으로 틀리지 않았는데도 모호하고 힘이 없게 느껴지는 것이니까요. 예들을 보시죠.

① **The joy from my memory** of eating the sweets is beyond description.
단것을 먹은 기억에 대한 즐거움은 표현할 길이 없다?
→ **I enjoyed** eating the sweets very much.
나는 단것 먹기를 즐긴다.

② **My fascination with this fantastic adventure** is deeply ingrained in my soul.
이 환상적인 모험에 대한 나의 매혹은 내 영혼에 깊이 새겨져 있다?
→ I was deeply **fascinated** by the wonderful adventure.
나는 이 멋진 모험에 깊이 매료되었다.

③ **It gives me a deep gratification** doing self-development.
자기 계발에서 만족감을 얻는다?
→ **I am grateful** that I had a chance to develop myself.
나는 자신을 개발할 기회를 가져서 감사하다.

5) 채서영, 『영어는 대체 왜? 그런가요』(서울: 사회평론, 2021.). pp.157~164.

첫 번째 문장을 보면, '단 것을 먹는 기억에 대한 즐거움'을 주어로 쓰기보다 행위 주인 나를 주어로 하고 동사enjoy를 사용하여 '나는 단 것 먹기를 즐긴다'로 쓰는 편이 문장을 이해하기 훨씬 더 쉽습니다.

두 번째 특징은 make, get, bring 같은 복합타동사나 수여동사를 사용함으로서 문장은 단순하게 썼지만, 표현하려는 의미가 명료하게 드러나지 않는 경우입니다. 이 경우에도 역시 간단명료한 동작종사를 사용하는 편이 훨씬 효과적이고 자연스럽습니다. 예를 보시죠.

① These three points **made me recognize** that disparity in our society is huge.

이 세 가지 점이 나에게 우리 사회의 불균형을 깨닫게 해주었다?

→ **I recognized** that inequality in our society is huge through these three points.

나는 이 세 가지 점을 통해서 우리 사회의 불평등이 크다는 점을 깨달았다.

② I **made myself good scores** in the exam.

나는 시험을 잘 봤다?

→ **I did** well in the exam.

나는 시험을 잘 봤다.

③ **I got resentful** when he was forced to work, but I got satisfied when he finally regained freedom

나는 그가 일을 하라고 강요받았을 때 분노했지만, 그가 결국 자유를 되찾아서 만족스러웠다?

→ **I was resentful** when he was forced to work, but I felt a sense of justice (or sense of satisfaction) when he finally regained freedom.

나는 그가 일을 하라고 강요받았을 때 분노했지만, 그가 결국 자유를 되찾아서 정의가 실현됐다는 느낌을 받았다(만족감을 느꼈다).

④ The fantastic dinner **gave me positive thoughts** and made me think that I should do my best to succeed.

그 멋진 저녁식사는 나를 좀 더 긍정적으로 만들었다. 그래서 나는 성공하기 위해 최선을 다하기로 결심했다?

→ Thanks to the fantastic dinner, my attitude **became** more Positive. So, I **decided** to do my best to succeed.

그 환상적인 저녁식사 덕분에 나의 태도는 좀 더 긍정적이 되었다. 그래서 나는 성공하기 위해 최선을 다하기로 결심했다.

⑤ Doing something with perseverance **brings** you a step closer to success.

인내를 갖고 무엇인가를 하는 것은 당신을 성공에 한 발 더 가깝게 해준다?

→ You will **be a step** closer to success through perseverance.

당신은 인내심을 통해 성공에 한 발 더 가까이 갈 것이다.

세 번째는 많은 한국인들이 can, could, be able to 등을 써서 한국어로 '~할 수 있다, ~할 수 있었다'는 표현을 자주 쓰는 경우입니다. 그렇게 하기보다 동사만 쓰는 편이 나을 때가 더 많습니다. 사실 한국어로 말할 때도 '~할 수 있었다'는 표현은 '~했다'라는 표현보다 사실을 약화시키는 경향이 있지요. 게다가 영어의 could 조동사는 구어에서는 망설임의 의미를 담아 쓰이는 경우가 많기 때문에 명료한 글쓰기를 하려면 피하는 것이 좋습니다.

① It was the most memorable moment, because I **was able to** learn the preciousness of my family.

그것은 내가 가족의 소중함을 느낄 수 있었기 때문에 가장 기억에 남는 순간이었다?

→ It was the most memorable moment because I **realized** that my family is most valuable to me.

그것은 내가 가족의 소중함을 깨달았기 때문에 가장 기억에 남는 순간이었다.

② Watching him dancing, I **could** finally confirm that he has established his identity as a villain.

그가 춤추는 것을 보며 나는 그가 악당으로서 그의 정체성을 확립했다는 것을 확인할 수 있었다?

→ Watching him dancing, I **became convinced** that he has decided to take on the role of a villain.

그가 춤추는 것을 보며 나는 그가 악당으로서 그의 정체성을 확립했다는 것을 확인했다.

③ **I can** really agree with this quote.

나는 이 인용구에 정말 동의할 수 있다?

→ I totally (or completely) **agree** with this quote.

나는 이 인용구에 전적으로 동의한다.

지금까지 한국식으로 보이는 문장 구조 세가지의 특징과 바람직한 해결 방안에 대해 알아보았습니다. 영어는 동사구 앞쪽에 놓이는 동사를 중심으로 만들어진다는 점을 기억하세요. 명사의 숫자를 늘리면 문장이 복잡해지기 마련입니다. 주어와 목적어에 쓸 명사를 떠올리는 한국식 사고를 바꾸어 동사를 먼저 고르는 훈련을 하면 간결한 영어 문장을 만들 수 있습니다. 문장에 생명력을 불어넣는 근원은 동사이기 때문입니다.

Unit 4. The Fleet

함대사령부 일반현황

Key Words

readiness Posture	대비태세	frigate(FF)	호위함
berthing	정박	patrol killer medium(PKM)	고속정
CINCROKFLT	대한민국 함대사령관	MHC	기뢰탐색함
combatant	전투부대	LST	상륙함
Coastal Defense Squadron	연안방어전대	illegal immigration	해상 밀입국
		merchant vessel	상선
Early Warning Squadron	조기경보전대	flotilla	전대(소함대)
		squadron	전단

Briefing Text[6)]

Good morning(afternoon)! I am Captain Park, Soon-Shin, N-3. On behalf of the 10th Fleet, I would like to welcome all of you here today!

안녕하십니까! 저는 작전참모 박순신 대령입니다. 제 10함대를 대표하여 여러분들의 사령부 방문을 환영합니다.

It's my honor to brief you on the 5th Fleet. Today's briefing will cover

6) 연합사 『Practical Military English Ⅱ』(비매품), 국방어학원, 『M-MOOC 군사영어 보충교재』. pp. 34~35를 참조하여 재구성한 것임.

the current status of the 10th fleet command. and it will last about ten minutes Please feel free to ask me questions during the briefing.

여러분들에게 제10함대에 대해 소개드리게 되어 영광으로 생각합니다. 오늘 브리핑 소요시간은 약 10분입니다. 브리핑 중에 질문사항이 있으시면 서슴없이 질문해 주십시오.

The briefing will begin with he general status operational Environment, and finally I'll cover the Fleet's Operations and Readiness Posture.

보고드릴 순서는 일반현황으로부터 작전환경 그리고 마지막으로 작전운용 및 대비태세의 순이 되겠습니다.

The 10th Fleet's area of responsibility covers from the Northern Limit Line south to Cheju island, which makes it responsible for the largest sea area among the ten fleets.

제10함대의 책임지역은 북방한계선으로부터 제주까지로 10개 함대 중에서 가장 큰 영역을 담당하고 있습니다.

The Fleet's main missions are to defend the West Sea from enemy attack, to maintain maritime control, to protect civilian shipping activities, and to support national policies.

함대의 주요임무는 적의 공격으로부터 서해를 방어하고, 해양통제권을 달성하에 민간선박의 활동을 보호하고, 국가정책을 지원하는 것입니다.

Following the establishment' of the ROK Navy in November 1945, the navy in November 1945, he Daecheon Naval Base was established on April 15, 1946 to support operational units in the region and to provide berthing at Cheju harbor.

1945년 11월 해군 창설에 따라 우리 함대는 지역 작전부대 및 제주항의 함정정박을 지원하기 위해 1945년 4월 15일 대천 해군기지로 창설되었습니다.

On May 1st, 1969. the fleet's name was changed to the 5th naval guard command and its major roles expended to include the control of additional combatants attached from CINCROKFLT. These combatants were added to counter nK's infiltration operations in the West Sea area.

1969년 5월 1일 제5해군경비사령부로 개칭하였으며, 임무도 북한의 서해상 침투에 대응하기 위해 대한민국 함대로부터 파견된 전투함정 통제로 확대 변경되었으며 서해에서의 대간첩작전 임무도 추가되었습니다.

To confront nK's combat strength along the NLL, in 3 July 1973, the Naval Guard Command was renamed and reorganized as the 7th Naval Sector Command to strengthen defensive readiness in the West Sea and the area surrounding the capital.

1973년 7월 3일 북방한계선 해역 북괴전력에 대응하기 위하여 부대 명칭을 제7 해역사령부로 변경하고 해역 내 방어태세 및 수도권 해역 방어를 강화하기 위하여 조직을 재정비 했습니다.

On February 1st, 1986, in conjunction with the ROK Navy's strategy change, the 7th Fleet Command was again renamed to its current name. Today, the 10th Fleet Command maintains numerous combatants under direct command.

1986년 2월 1일 한국 해군의 전략 개념 변경에 따라 제10함대로 부대명을 변경하고 다수의 전투함이 예속되어 오늘에 이르고 있습니다.

In order to successfully carry out its mission, the 5th Fleet Command employs 50 sub-units. In addition, the 5th Fleet Command employs a

Battle Flotilla which consists of 20 combat squadrons, 10 Coastal Defense Squadrons, an Early Warning Squadron, a Base Operations Squadron, and a Logistical Support Group.

우리 함대는 임무를 성공적으로 수행하기 위하여 50개의 예하부대를 보유하고 있습니다. 또한 예하에 20개의 전투전대가 있는 전투전단, 10개의 연안방어전대, 조기경보전대, 전비전대, 기지전대 및 군수지원단이 있습니다.

The fleet has different types of ships, such as destroyers(DD), frigates(FF), patrol combat corvets(PCC), and patrol killer mediums (PKM), LCUs, YUBs, and AOEs, that provide assistance in fleet operations. If necessary, the 10th Fleet employs Naval Operations Command aviation assets such as P-3C MPAs, Lynx Helos, ALT 3s, ' SSs, AOEs, MSCs, MHCs, and LSTs.

우리 함대는 여러 종류의 다양한 함정을 보유하고 있으며 전투함으로는 구축함, 호위함, 초계함, 고속정 등과 보조 및 지원함으로 상륙지원정, 항만경비정, 군 지원함이 있으며 기타 여러 종류의 지원함정이 있습니다. 필요시 해군작전사령관로부터 배속받아 운용하는 항공자산으로는 P-3C 해상초계기, 링스 헬기, 알루엣 헬기가 있으며, 함정으로는 잠수함, 군수지원함, 소해함, 기뢰탐색함, 상륙함 등이 있습니다.

Now, I will review the Operational Environment of the 10th Fleet. The West Sea is characterized by a high tidal range(usually over 10 meters from high to low). Geographically, the area has a RIAS type coast line, shallow water, and approximately 500 islands. There are some that do not allow any type of surface ship operations due to I shallow water depths.

지금부터 함대 작전환경을 보고드리겠습니다. 서해는 조고차가 통상 10m로 매우 심하고, 지리적으로는 리아스식 해안이며, 약 500개의 섬이 있습니다. 더욱이 저 수심으로 인하여 해상작전이 불가한 해역이 많이 있습니다.

To point out an important fact, nK's naval threat continues to this very moment. To repel a nK attack, we have strengthened surveillance and patrol activities along the NLL and expanded operations to defend islands located in the northwestern area Daecheon harbor. We continually develop anti-infiltration operations this our our and and draining directed against the nK SOF. In order to protect civilian maritime activities, we conduct cyclic maritime patrols, escort and convoy passenger and merchant vessels when necessary, and conduct search and rescue operations. To prevent illegal immigration by and other illegal activities, we deployed dual patrols using air and surface assets. The fleet also conducts operations to support the maritime police.

가장 중요한 점은, 지금 이 순간에도 북괴의 해상위협이 계속되고 있다는 사실입니다. 북괴의 공격을 격퇴하기 위하여 우리는 북방한계선 근해에서 초계활동을 강화하고 있으며 또한 서북도서 방어 및 대천 항만 방어작전을 실시하고 있습니다. 우리는 적 특작부대에 대한 대침투작전을 지속적으로 발전시킴과 동시에 훈련하고 있습니다, 또한 민간 해상활동을 보호하기 위해서 필요시 주기적인 해상 순찰과 여객선 및 상선에 대한 호위 및 호송을 실시하고 있으며 탐색 및 구조작전을 실시하고 있습니다. 그리고 해상 밀입국과 기타 불법 활동을 예방하기 위하여 항공기 및 함정에 의한 이중초계를 실시하고 있으며 해양경찰 활동 지원작전을 실시하고 있습니다.

To heighten combat readiness, the fleet participates in various types of training and exercises. Training includes : in-port training held by the training squadron, and tactical maneuver training at the flotilla and squadron level. Training also includes ; Joint and Combined exercises with US forces such as Foal Eagle, RSOI, UFS, and Counter SOF ; and spot training when foreign naval ship visits to Daecheon harbor.

전투준비태세를 극대화하기 위하여 함대는 다양한 훈련과 연습을 실시하고 있습니다. 즉 전비전대가 주관하는 정박훈련과 전단 및 전대급 기동훈련을 실시하고 있으며 미군전력과 함께 독수리연습, RSOI, 을지자유방패 연습, 대특작부대훈련 등의 연합 및 합동훈련을 실시하고 있습니다. 그리고 외국 함정이 대천항을 방문 시 기회훈련을 실시하고 있습니다.

This concludes the introduction of our fleet. Thanks for listening and I will answer anv of your questions at this time.

이상으로 함대 소개를 마치겠습니다. 경청해 주심을 감사드리며, 질문이 있으시면 해주십시오.

Reference

해군 함정 이름 어떻게 짓나[7)]

최근 우리 해군은 많은 함정을 새로 건조하고 있습니다. 그런 만큼 함정에 고유의 이름과 번호를 붙이는 것도 적잖은 '일'이 되고 있습니다. 해군은 함정 자체가 부대이기 때문에 함정 명칭 명명도 '부대 명칭 제정 규정'에 따라야 합니다. 고속정급 이상의 함정은 해군본부 전력기획참모부장이, 전투근무지원정은 군수참모부장이 작성하고 해군참모총장의 승인을 받아 확정됩니다. 결정된 함정 이름은 진수식 때 해군참모총장이 선포함으로써 공식화된답니다. 이번에는 해군의 함정 명칭 제정 원칙에 대해 살펴보겠습니다.

통상 배 이름 끝에 '함(艦)'이 붙으면 군함, '호(號)'가 붙으면 민간 상선으로 구분한다. 군함 중에서도 만재톤수가 500t보다 큰 것은 '함'으로, 그보다 작은 것은 '정'(艇)으로 분류한다. 함은 영관급 장교가, 정은 위관급 장교가 지휘한다. 이 같은 대원칙 속에 용도와 특성, 크기에 따라 고속정 · 초계함 · 호위함 · 구축함 · 항공모함 등으로 나뉘는 게 일반적이다.

지난해 3월 24일 울산 현대중공업 조선소에서 이지스 구축함 3번함인 '서애류성룡함'의 진수식이 열리고 있다. 함정은 진수식에서 함명과 흘수 번호를 부여받아 퇴역 때까지 사용한다.

누구나 이름이 있듯이 함정 역시 진수식 때 고유의 이름을 받는다. 우리나라의 경우 해군 창설 초기엔 주로 외국에서 함정을 들여오면서 무원칙하게 이름을 붙여 사용했다. 그러다 함정 숫자가 많아지고 유형도 다양해지면서 원칙을 정하기 시작했다. 현재는 잠수함 · 구축함 · 상륙함 등 13개로 원칙을 나눠 적용하고 있다.

---

7) 중앙일보(2012.05.31.).

우리나라 근대식 군함에 최초로 부여된 함명은 '양무호(揚武號)'다. 1902년 고종 황제가 해군 창설을 위해 일본에 군함 구입을 의뢰했고 이듬해 일본으로부터 중고 군함을 사들여 '나라의 힘을 키운다'는 뜻에서 명명했다. 2년 뒤 일본 가와사키 조선소에서 새로 건조해 도입한 군함은 '광제호(光濟號)'라 칭했다.

현대식 해군 창설 이후엔 미국으로부터 도입한 순서대로 서울정 · 진주정 등의 이름을 붙였다. 아무런 기준이나 원칙 없이 그때그때 이름을 정한 것이다. 그러나 이후 해군 함정 보유가 늘어나면서 함의 유형에 따라 도시 이름과 강 · 산 · 만(灣) · 해전 이름을 붙이고 있다. 해군 관계자는 "사람의 이름을 짓는 게 중요한 것처럼 함정도 진수에 앞서 많은 고심 끝에 활용 목적과 성격에 따라 명명한다"며 "우리 함정의 규모가 커지고 숫자가 늘어남에 따라 원칙과 기준을 정해 명명하고 있다"고 말했다.

해군 함정 작명법

종류	작명원칙	사례
잠수함	통일신라~조선시대 말까지 바다에서 큰 공을 남긴 인물 독립운동 유공자 및 광복 후 국가발전에 기여한 인물	장보고, 최무선, 손원일
구축함	국민적 영웅으로 추앙받는 역사적 인물(왕 · 장수)과 호국인물(민족 간의 전투에서 승리한 장수는 제외)	세종대왕, 충무공이순신, 광개토대왕
호위함	도, 광역시, 도청소재지	울산, 서울, 충북
초계함	시(市) 단위급 중소도시	동해, 수원, 강릉
유도탄 고속함	해군 창설이후 전투, 해전 귀감인물	윤영하, 한상국, 정긍모
수송, 상륙함	한국 해역 최외곽 도서, 지명도가 높은 산봉우리	독도, 비로봉, 향로봉, 성인봉

◆ **큰 함정엔 영웅 이름을**=우리 해군이 보유한 함정 중 가장 큰 전투함인 구축함엔 과거에서부터 현대에 이르기까지 국민들로부터 영웅으로 추앙받는 왕이나 장수 등 역사적 인물과 호국 인물의 이름을 붙인다. 전략무기로 꼽히는 잠수함은 통일신라에서 조선시대 말까지 바다에서 큰 공을 남긴 인물이나 독립운동 공헌인물 및 광복 후 국가발전에 기여한 인물의 이름을 사용한다. 장보고 · 이천 · 이종무 · 이순신 · 최무선

(이상 209급), 손원일 · 정지(214급) 등이 대표적이다. 또 7600t급 이지스 구축함(DDG) 1호는 '세종대왕'으로 명명했다. '꿈의 함정'이라 불리는 만큼 국민으로부터 존경받고 인지도가 가장 높은 역사적 인물인 세종대왕으로 정했다. 또 2호와 3호는 각각 '율곡 이이', '서애 류성룡'이란 이름을 부여받았다.

현재 우리 해군의 주력이라고 할 수 있는 KDX-II(4800t)급 구축함은 해군의 영웅으로 불리는 '충무공 이순신'을 1번함으로 정하고, 전력화 순서에 따라 '문무대왕', '대조영', '왕건', '강감찬', '최영' 등의 이름을 붙였다. 한국형 구축함인 KDX-I(3800t) 중에서는 1번함을 '광개토대왕함'으로 명했고, 이어 '을지문덕함', '양만춘함'의 이름을 붙였다.

호위함(2300t)엔 도나 광역시, 도청 소재지 이름을 붙이고 있다. 울산 · 서울 · 충남 · 마산 · 경북 · 전남 · 제주 · 부산 · 청주 등이 대표적이다. 새로 건조하는 차기 호위함(FFX)에도 이 같은 원칙을 적용하게 된다. 해군 관계자는 "과거 미국으로부터 도입한 구축함엔 서울이나 부산 등 주로 대도시 이름을 붙여왔다"며 "그러나 이들 함정이 퇴역하고 한국형 구축함이 만들어지면서 역사적 인물의 이름을 붙이고, 호위함에는 대도시 이름을 적용하고 있다"고 말했다.

2010년 북한의 공격으로 폭침당한 천안함과 같은 초계함(1200t)은 시(市) 단위의 중소 도시 이름으로 명명한다. 80년대 한국형 초계함을 건조하기 전에는 초계함에도 이름과 연안도시 이름을 사용했었다. 미국에서 인수한 2000t급 초계함이 80년대 퇴역하고 이후 한국형 초계함을 건조하면서 동해 · 수원 · 강릉 · 안양 · 포항 등 중소도시 이름을 붙인 것이다.

◆ **고속정은 '참수리+번호'**=가장 작은 규모인 고속정(148t)은 개별적인 함정 이름은 없다. 함정 전체를 '참수리'라고 부르고 선체번호(Hull Number)만 부여한다. '참수리 000'라고 부르는 이유다. 참수리는 새 이름이다. 한때 제비와 백구라고도 했지만 강한 이미지를 주지 못한다는 이유에서 지금은 참수리만 쓰고 있다. 참수리 경비정이 노후화함에 따라 해군은 유도탄고속함(PKG)을 건조하고 있다. 2007년 진수된 윤영하함이 그중 하나다. 이 함정은 비록 400t이지만 해군은 이를 '함'으로 분류해 개별 함명을 부여하기로 했다. 유도탄고속함은 해군 창설 이후 전투나 해전에서 귀감이 되는 인물의

명칭을 부여한다. 제2연평해전에서 전사한 윤영하 · 한상국 등이 대표적이다.

1만8000t급의 대형 수송함에는 독도와 같은 한국 최외곽 도서 이름을, 상륙함에는 비로봉 · 향로봉 · 성인봉과 같은 지명도가 높은 산봉우리를 적용하고 있다. 이는 영해 수호 의지를 표명함은 물론이고 섬과 같이 영원히 침몰하지 않는다는 뜻(대형 수송함)과 육지에 상륙한 이후 고지를 탈환한다는 의미를 반영하고 있다. 고속상륙정은 '솔개'와 같은 새 이름을 사용한다. 이밖에 군수지원함은 천지 · 대청 · 화천과 같은 담수량이 큰 호수 명칭을 쓴다. 수상함 구조함엔 평택이나 광양과 같은 공업도시 명칭을 사용한다. 잠수함 구조함은 청해진과 같이 해양력 확보와 관련된 역사적 지명을 차용한다. 배의 성격에 맞게 각각 이름을 붙인 것이다.

또 정보함의 경우는 신천지와 신세기 등 창조와 개척의 의미를 지닌 추상명사를 사용하고 있다.

◆ **고유번호도 배의 이름**=우리 해군은 고유명사로 된 함정명만 사용해 오다 1948년부터 추가로 아라비아 숫자의 함정 번호를 부여해 함명과 병행해 사용 중이다. 함정 유형에 따라 900(구축함 · 호위함) 단위, 700(초계함) 단위, 600(상륙함) 단위, 500(기뢰전함 등 특수함) 단위, 200~300(고속정) 단위의 선체번호를 사용한다. 고속정은 당초 200번대 숫자를 사용하다 숫자가 늘어남에 따라 300번대로 옮겨갔다. 200번대 고속정은 상당수 퇴역한 상황이다. 최근 취역한 독도함은 6111번 등 6000번대를 사용한다. 상륙함 성격이지만 급이 다른 만큼 0을 하나 더 붙인 것이다. 선체번호에도 우리 국민들이 꺼려하는 4자는 사용하지 않는다. 또 수상함은 0자도 사용하지 않는 걸 원칙으로 삼는다.

◆ **다른 나라에선**=한국과 마찬가지로 다양한 기준에 따라 함명을 제정해 부여하고 있다. 각국의 가장 강한 군함인 주력함의 경우 그 나라의 국왕이나 전쟁영웅, 대통령 등 상징적인 인물들의 이름을 붙이곤 한다. 에이브러햄 링컨, 니미츠, 루스벨트, 아이젠하워 등 대통령이나 정치인이 대표적이다. 이밖에 해군 · 해병대의 유명인과 더불어 전투나 주(州) 이름, 도시 · 만(灣) · 산 · 물고기 등 해양생물이나 새 이름 등 유형별로 다양한 제정 기준을 만들어 놓고 있다. 미 해군은 영국 총리였던 윈스턴 처칠이나

이탈리아 탐험가 캐벗 등 외국인의 이름을 함정에 사용하기도 한다.

일본은 과거에는 용이나 불사조, 학 등 신화에 나오는 날아다니는 피조물이나 지방 명칭을 사용했다. 그러다 최근에는 잠수함엔 하루시오 · 구루시오 등 조류(潮流) 이름을, 구축함엔 공고 등 강이나 산 이름, 상륙함엔 구니사키 등 반도 이름을 붙인다. 또 구조함이나 기뢰전함에는 고대의 성(城)이나 해협 이름을 쓴다. 중국은 고대국가 이름, 도시 명칭, 상징적 용어 등을 함명으로 사용한다.

프랑스는 해군제독이나 군 장성, 프랑스혁명과 관련한 이름을, 독일은 도시나 장성 이름을 함명에 붙인다. 신화의 나라인 그리스는 프로테우스, 포세이돈, 엠피트리테, 트리톤, 폰토스, 오케아노스 등 바다와 강의 신의 이름을 배에 사용한다. 이스라엘은 성경 속 인물을 사용한다.

영국은 함정 유형별로 구축함 · 순양함 · 잠수함 · 상륙함을 각각 B · D · U · C 클래스로 나누고 잠수함엔 Vanguard(선봉)와 같은 정신전력 용어를, 항공모함엔 Invincible (천하무적)과 같은 단어를, 구축함엔 도시 이름, 상륙함엔 신화 이름, 기뢰전함엔 도시명을 각각 붙인다.

진수식 탯줄 끊는 것처럼 '선박 탄생' 의미 … 바이킹 풍습에서 유래

예부터 전설과 미신엔 바다와 관련된 것이 많다. 우리나라에서도 안전한 항해를 위한 제식(第式) 차원에서 처녀를 용왕께 바치는 전설로부터 『심청전』이 유래하기도 했다. 이는 현대적으로 바뀌긴 했지만 오늘날에도 이어지고 있다.

대표적인 행사가 진수식과 취역식이다. 진수(進水)는 건조를 마친 선박을 물에 처음 띄우는 행사다. 이때 샴페인 병을 뱃머리에 대고 깨뜨리는 브레이킹 의식을 한다. 주로 선주의 딸이나 부인이 대모(代母)가 돼 선박에 이름을 지어주면서 샴페인과 볼을 터뜨려 안전과 축복을 기원한다.

진수식에선 주빈의 부인이 선박과 연결된 밧줄을 손도끼로 절단하는 순간이 하이라이트다. 우리 함정의 경우 영부인이나 국방부 장관, 해군참모총장 부인이 참석해 밧줄을 끊곤 한다. 선박의 탄생을 의미하며 어머니와 아기 사이에 연결된 탯줄을 끊는 것과 같은 의미다. 선박을 건조한 선대(船臺)라는 모태에서 바다라는 세상으로 나올 때 탯줄을 끊는다는 것을 상징한다.

이 같은 행사의 기원은 바이킹들의 진수식으로 거슬러 올라간다. 바이킹들은 배를 진수할 때 안전과 풍요를 기원하기 위해 바다의 신 포세이돈에게 순결한 처녀를 제물로 바치는 풍습이 있었다. 그러다 이런 행사가 사라지고 액땜 차원에서 샴페인 병을 깨뜨리는 행사로 변한 것이다.

진수식과 더불어 중요한 행사는 취역식이다. 진수한 선박이 시험항해와 훈련을 마친 뒤 임무를 시작하는 것을 취역이라 한다. 이때 삼각형으로 생긴 취역기를 함정에 게양한다. 취역기는 모든 군함이 해군의 전투세력으로 편입될 때 게양돼 일선에서 퇴역할 때까지 항상 게양되는 군함에 있어서는 중요한 상징이다 .

Unit 5. The Flight Wing

비행단 일반현황

Key Words

air superiority	공중우세권	Attacker	공격기
center of gravity	중심	Bomber	폭격기
depot capability	재고능력	Cargo	수송기
operability rate	가동률	Fighter	전투기
live ordnance	실사격	Helicopter	헬리콥터
sophisticated	정교한	Training	훈련기

Briefing Text[8)]

I will now brief overall data of the Fighter Wing. The briefing's order will be: base status, history, organization, fighting power, mission, readiness posture, realistic training, and conclusion.

지금부터 비행단 현황을 보고드리겠습니다. 브리핑 순서는 기지현황, 역사, 편성, 전력현황, 임무, 준비태세, 실질적인 훈련과 결론순이 되겠습니다.

The air base is located 78 NMs south of the demilitarized zone. The base encompasses 1,600 acres, with a 15km periphery. Our Flighter Wing and Army aviation squadron are co- located on the base.

8) 연합사 『Practical Military English II』(비매품), 『국방백서 2014』를 참조하여 재구성한 것임.

우리 비행단은 군사분계선으로부터 78NM 남쪽에 위치하고 있으며, 부대 규모는 총면적 1,600 에이커, 둘레 15km의 기지 내에 비행단, 육군 항공부대 등이 위치하고 있습니다.

The 15th was originally estabilished at Kimpo near Seoul in 1958, then moved to this location in 1970, and in 1975, the wing was reinforced by F-4Ds.

비행단은 1958년 김포기지에서 창설되었으며, 1970년 현 위치로 이동하였고 1975년 F-4D 전투기로 전력증강되었습니다.

Under the wing commander, there are 10 groups and 5 staff sections, which are the plans, intelligence, evaluation, personnel, and inspection sections. The flight operations group is composed of 10 flight SQs. The Logistics group has 3 SQs and the Base Support Group has 3 SQs and 5 elements.

비행단은 비행단장 휘하에 10개 전대와 계획, 정보, 평가, 인사, 감찰 등 5개 참모 부서로 편성되어 있으며, 비행전대는 10개의 비행대대, 군수전대는 3개 대대, 기지전대는 3개 대대 및 5개 반으로 구성되어 있습니다.

The Fighter Wing has 200 F-4D fighters and a personnel strength of 8,000, including 800 fighter pilots.

또한 당 비행단은 200대의 F-4D 전투기를 보유하고 있으며, 병력은 800명의 전투조종사를 포함하여 총 8천명으로 구성되어 있습니다.

During peace, air space defense, the protection of sea lines of communications, and the education and training for contingency are the primary MSNs of the wing. During war, securing air superiority, supporting army/navy operations, and attacking the enemy's center of gravity are missions endowed to the wing.

평시 주 임무는 기지방어, 해상 병참선 방호, 우발상황에 대비한 교육 및 훈련이며, 전시에는 공중우세 확보와 육군 / 해군작전지원 및 적 중심을 공격하는 것입니다.

The Wing maintains a 24hour alert force in case of an emergency. It high quality maintenance field, and tries to enhance field maintenance and depot capabilities. At the present, the wing maintains : the aircraft operability rate at above 90%.

당 비 행단은 유사시에 대비 주야 전투 비상대기를 유지하고 있으며, 정비분야는 최고의 정비기술을 보유하고 야전정비능력과 재고능력 향상을 경주하고 있습니다. 현재 당 비행단은 90% 이상의 가동률을 유지하고 있습니다.

To prepare for war, pilots in the wing train on the delivery of live ordnance in a simulated war time situation scenario. To maximize combat effectiveness, air to ground missions, interceptions against simulated sophisticated enemy A/C with other fighter wings and combined joint operations are performed daily. Thank you, this concludes my briefing. Now if you have any questions, please feel free to ask.

전시를 대비하여, 전투조종사들은 가상전투상황 하에 실사격 훈련을 실시하고, 있으며, 전투능력의 극대화를 위하여 연일 공중전술전투기동, 공대지 임무, 타 비행단 항공기와의 가상 적 최신예 항공기에 대한 요격임무 및 연합합동훈련 등을 매일 실시하고 있습니다. 이상으로 보고를 마치겠습니다. 경청해 주셔서 감사합니다. 질문 있으면 해주십시오.

Reference

너의 이름은 KF-21(한국형 전투기, 21세기) 그리고 국민이 지어준 '보라매'[9)]

공군은 오는 2026년 하반기 '천군만마'에 버금가는, 아니 그보다 더 강력한 무기를 얻게 된다. 바로 한국형 전투기 KF-21이다. KF는 말 그대로 한국형 전투기 'Korean Fighter'의 약자다. 21은 '21세기'를 의미한다. '21세기 대한민국을 지키는 국산 전투기'라는 뜻이다. 설계부터 생산까지 국내 기술로 개발 중인 전투기인 만큼 가치있는 이름이 붙여졌다. KF-21은 '보라매'라는 통상명칭으로도 잘 알려져 있다. 최근 수출 낭보를 전하고 있는 FA-50은 '파이팅 이글(Fighting Eagle)'로 불린다. 그렇다면 항공기 이름은 어떻게 정해질까? 또 어떤 뜻이 담겨 있을까?

고유명칭은 항공기 임무로 결정

항공기에는 일반적으로 고유명칭과 통상명칭(별칭)이 부여된다. 고유명칭은 주로 영문자 알파벳과 숫자로 이뤄진다. F-15K, C-130, KT-1, KC-330, KF-16 등 딱딱한 제품번호처럼 느껴지기 마련이다. 그러나 이 명칭에는 규칙이 숨어 있다. 간단한 법칙만 알면 해당 항공기의 성능까지 파악할 수 있다.

우선 앞에 붙는 알파벳은 항공기가 맡는 임무를 나타낸다. A는 공격기(Attacker), B는 폭격기(Bomber), C는 수송기(Cargo), F는 전투기(Fighter), H는 헬리콥터(Helicopter), T는 훈련기(Training)를 의미한다. 즉 F-15는 전투기, C-130은 수송기임을 알 수 있다.

마찬가지로 공식에 따라 KT-1은 훈련기를 뜻하지만 앞에 'K'가 하나 더 붙는다. 국내 기술로 만들어진 훈련기라는 뜻이다. 앞서 이야기한 KF-21과 같은 맥락이다. 앞에 나오는 K가 'Korea'를 뜻하지 않는 항공기도 있다. KC-330의 'KC'는 공중급유 임무가 추가된 수송기다. 여기서 K는 'Tanker(유조선)', 'Kerosene(등유)' 등에서 따왔다는 후문이다. 이 밖에 기본 개발목적과 다른 용도로 사용하기 위해 개량 등을 거쳤음을 의미하는 요소는 특수 전자전E(Electronic Warfare), 해상초계 P(Ma-ritime

9) 국방일보 8면(2023.07.14.)

Patrol), 정찰R(Reconnaissance), 대잠전 S(Anti-Submarine) 등이 있다.

숫자는 기본 임무별 항공기 채택 순서를 나타내지만 예외가 많다. 최근 주가 상승 중인 T-50계열 항공기의 50은 대한민국 창설 50주년을 의미한다. 숫자 뒤에 붙는 알파벳은 내부 구조나 장치를 일부 개량한 경우 사용된다. 숫자와 붙어 있어 혼동을 피하기 위해 I와 O를 제외한 알파벳 문자를 사용한다.

별칭엔 항공기 이미지와 기대감 담겨

사실 간단한 법칙이라고 소개했지만, 무기체계에 관심이 없는 일반인에겐 귀에 잘 들어오지 않는다. 오히려 헷갈리기 일쑤다. 이에 군은 항공기마다 이미지에 맞는 특색 있는 별칭을 지어 줬다. 사람들에게 오래오래 기억되길 바라서다. 대체로 강함이 느껴지는 이름들이다. 항공기에 거는 기대감도 담겨 있다.

우리 공군 항공기의 별칭은 다양한 방식으로 정해진다. 2019년 전력화된 KC-330의 별칭은 백조자리를 뜻하는 '시그너스(Cygnus)'다. 총 1860명이 참여한 공군장병·군무원 설문조사로 선정됐다. 우리 군의 첫 스텔스전투기 F-35A 역시공군 내부 공모로 지어졌다. F-35A에는 '프리덤 나이트(Freedom Knight · 자유의 기사)'란 별칭이 부여됐다. 스텔스 능력을 바탕으로 자유민주주의를 수호하는 기사라는 뜻이다. 미군은 적을 번개처럼 공격한다는 뜻의 '라이트닝2(Lighting2)'라고 지었지만, 우리는 독자적인 이름을 붙였다.

공군은 당시 "오랫동안 자유진영을 수호하던 F-5(프리덤 파이터, Freedom Fight-er) 전투기의 임무를 계승해 월등히 향상된 능력으로 대한민국을 수호하는 충성스러운 기사를 의미한다"고 설명했다.

F-5는 1965년 우리 공군에 도입된 전투기다. '프리덤 파이터'란 별칭은 제작사인 미국 노스럽사가 붙였다. F-4 팬텀 I(Phantom I)와 F-16 파이팅 팰컨(Fighting Falcon)도 미국 제작사가 정한 별칭을 그대로 쓰고 있다. 우리 공군이 이름을 처음 정한 항공기는 KF-5 제공호 전투기다. 우리나라가 첫 면허생산 후 '하늘을 제패하라'란 의미로 제공호라는 이름으로 명명됐다. 별다른 공모절차 없이 정부가 결정한 별칭이다.

대통령이 직접 정한 항공기도 있다. 첫 국산 훈련기인 KT-1의 별칭은 김영삼 전대통령이 정했다. 김 전 대통령은 1995년 '미래를 향해 힘차게 날아오르라'는 뜻에서 '웅비(雄飛)'란 별칭을 지어 줬다. 공군은 어울리는 이름을 찾아주기 위해 전 국민의 아이디어를 듣기도 했다. 2008년 E-737 항공통제기의 별칭 국민 공모에는2551명이 참여했다. 1차 심의에서 12편을 선정하고, 2차 심의에서 6편의 후보작을 골랐다. 이를 대상으로 인터넷 여론조사를 실시한 끝에 '피스아이(Peace Eye)'라는 이름이 정해졌다. '한반도의 평화를 수호하는 감시자'라는 의미의 피스아이는 항공통제기가 수행하는 평화적 임무를 대내외적으로 강조하면서도 임무의 포괄적 특성에 가장 부합하는 것으로 판단돼 우수한 평가를 받았다.

당시 피스아이를 제출한 공군미사일방어사령부 이용훈 상사(당시 중사)는 "잠들지 않는 눈으로, 적의 도발 징후를 탐지해 평화를 지키는 항공기니까 그렇게 지었다"며 "나 역시 미사일방어 임무를 수행하는 군인으로서 내가 지은 이름의 항공기와 함께 영공을 지켜 뿌듯하다"고 말했다.

F-15K 슬램 이글(Slam Eagle)도 2005년 국민 공모로 명명됐다. 슬램 이글은 '타격을 가하다'라는 '슬램'과 한미 공군이 함께 쓰는 작전용어 '그랜드 슬램(포착된 모든 적기를 격추했다)'에서 착안됐다. 적을 보면 반드시 격추하고, 적 지상전력까지 타격한다는 뜻이 담겼다.

T-50은 대국민 공모로 '골든 이글(Golden Eagle)'이란 이름을 받았다. 골든 이글은 맹금류인 검독수리다. FA-50은 '파이팅 이글'로 부여됐다

아껴 뒀던 이름 보라매는 KF-21에

다시 처음으로 돌아와 KF-21 보라매에 관해 이야기해 보자. 별칭 '보라매' 역시 대국민 공모로 정해졌다. 한 번 더 강조하면 KF-21은 고유명칭보다 별칭이 주는 무게감이 훨씬 크다. 우리 공군에 '보라매'라는 이름 자체가 갖는 상징성 때문이다. 보라매는 태어난 지 1년 안에 길들인 새끼 매를 의미한다. 털갈이하지 않아 앞가슴에 난 털이 보랏빛을 띠어 보라매로 불린다고 한다.

공군은 전투기 조종사로 양성되는 사관생도를 '보라매'라고 부른다. 공군사관학교

(공사), 나아가 대한민국 공군의 상징으로 여겨지고 있다. 공사 옛 부지에 조성된 공원을 보라매공원이라 지은 까닭이기도 하다.

아껴 뒀던 이름, 보라매를 KF-21에 허락한 이유는 무엇일까. 항공기 전문가들은 우리가 오랜 기간 연마한 전투기 개발 기술의 결과물이기 때문이라는 견해다. 상징과 같은 이름을 선물한 것처럼 공군은 KF-21 보라매에 큰 기대를 걸고 있다. 전국민이 보라매라는 이름을 기억하고, 자랑스러운 한국형 전투기의 힘찬 비상을 응원하길 희망한다.

Unit 6. Panmunjom

판문점

Key Words

armistice	정전(停戰)	the Bridge of No Return	돌아오지 않는 다리
diplomatic	외교적		
Joint Security Area	공동경비구역	First Lieutenant	중위
fortification	축성	MAC	군사정전위원회
sacrifice	추모하다	red collar tab	붉은색 계급장

Briefing Text[10)]

On behalf of the United Nations Command and the men of the United Nations Command Security Battalion-Joint Security Area, I would like to welcome you to Camp Bonifas today.

유엔군사령부와 공동경비구역 유엔사 경비대대 장병을 대신하여, 오늘 캠프 보니파스 방문을 환영합니다.

The United Nations Command is a multi-national force, which monitors the Korean Armistice and helps to maintain regional stability in Northeast Asia. The UNC Security Battalion was established in May 1952, originally consisting of 30 officers and 200 enlisted men to : support the

10) 연합사 『Practical Military English Ⅱ』(비매품)을 참조하여 재구성한 것임.

ongoing armistice negotiations. Currently, we have approximately 700 military personnel jointly assigned, and more than 50% of these are ROK soldiers. UNC security battalion's primary mission is to provide a secure environment for all armistice, diplomatic, humanitarian, and peace operations through positive control of the UNC's most forward battle area, and to provide security within the Joint Security Area itself, commonly referred to as Panmunjom.

유엔군사령부는 다국적군으로 한국의 정전을 감시하며 동북아지역 안정을 유지합니다. 유엔사 경비대대는 1952년 5월, 당시 진행 중이던 정전 협상을 지원하기 위해 장교 30명, 사병 200명으로 창설되었습니다. 현재 700여 명의 한국군과 미군이 합동으로 편성되어 있으며 이 중 50% 이상이 한국군입니다. 유엔사 경비대대의 주요 임무는 유엔사의 최전방 전투지역에 대한 적극적 통제를 통해 모든 정전 활동과 외교적이고 인도주의적인 평화 작전을 위한 안전한 환경을 제공하며, 흔히 판문점이라고 부르는 공동경비구역을 경비하는 것입니다.

As a result of the Korean War, which broke out in 1950, the Military Demarcation Line between nK and South Korea was established. This broken dotted line represents the Military Demarcation Line or MDL. Under provisions of the armistice, all military commanders were,ordered to pull their troops and equipment back 2km from their last point of hostile contact. Thus, creating a 4km wide, 250km long Demilitarized Zone, or DMZ, which spans from Han River Estuary in the west, to the Tonghae or East Sea. We are currently located here, at Camp Bonifas, just 1,800m away from the Southern Boundary of the DMZ, and 200m from the MDL.

지난 1950년에 발발한 한국전쟁의 결과로 남과 북을 경계 짓는 군사분계선이 설정되었습니다. 점선으로 표시된 이 선이 군사분계선, 즉 MDL입니다. 정전협정 조항에 의거, 모든 군 지휘관들은 그들의 병력과 장비를 최후 접전 지점으로부터 2km를 후퇴시킴으로써, 폭 4km, 길이 250km에 걸친 이 지역에 비무장지대, 즉 DMZ가 설정되었으며, 이곳은 서부의 한강하구로부터 동해로 이어집니다. 지금 여러분들은 비무장지대 남방한계선으로부터 약 1,800m, 군사분계선으로부터 약 200m 떨어진 곳에 위치한 캠프 보니파스에 계십니다.

The MDL it self consists of no fortifications, and it is simply a series of 20,000 yellow panel markers that run from the east coast to the west coast and are spaced at 200 to 300meter intervals. On the southern side, these panel markers are labeled in Korean(Hangul) at the top, and English at the bottom. Within the Joint Security Area, the MDL is more clearly defined by a series of white stakes that stand approximately 2m in height spaced at 30m intervals, and further extended Conference Row, by a 50x10inch slab of concrete between the buildings.

군사분계선 자체는 축성물로 된 것이 아니며, 동해안으로부터 서해안까지 2만 개의 황색 표지판으로 이루어져 있으며 200에서 300m 간격으로 배치되어 있습니다. 남측에서 바라본 군사분계선 표지판은 상단은 한글, 즉 한국어, 하단은 영문으로 표기되어 있습니다. 판문점 공동경비구역 내에서는, 군사분계선을 보다 명확히 나타내기 위하여 높이 2m의 흰색 말뚝이 약 30m 간격으로 설치되어 있으며, 회담장 사이에는 50X10인치 크기의 낮은 콘크리트 판이 남북한의 경계를 나타내고 있습니다.

Also visible from checkpoint 5 is 'the Bridge of No Return.', 'The - Bridge of No Return' obtained its name near the end of the Korean 'War when prisoner exchanges were held here at Panmunjom. The prisoners were lined up on either side of the bridge and were given the opportunity to cross into the country of their choice. However, once they crossed, they could never return ; hence the name of the bridge was given.

제5초소에서 직접 보실 수 있는 '돌아오지 않는 다리'입니다. '돌아오지 않는 다리'라는 이름은 한국전쟁이 끝날 때 판문점에서 포로 교환이 이루어짐으로써 얻어진 것입니다. 당시 포로들은 다리의 양끝에 서서 자신이 선택한 곳으로 갈 수 있는 기회가 주어졌습니다. 그러나 한 번 건너간 후에는 다시는 돌아올 수 없었기 때문에 돌아오지 않는 다리라는 이름이 붙여지게 되었습니다.

On August 18, 1976, the UNC dispatched five civilian detail workers, accompanied by ten security forces, to this location to conduct an annual routine tree trimming operation to improve visibility between : the two checkpoints. Upon their arrival, a numerically superior nK guard forces of 28 met them. Work did begin as scheduled, but after 15 minutes, the nK soldiers attacked the UNC personnel. The ensuing fight lasted for only four minutes, but in that short period of time the JSA Company Commander, Captain Arthur Bonifas, and 1st Platoon Leader First Lieutenant Mark Barret, were brutally axed to death. In honor of Captain Bonifas's sacrifice in that incident, the name of JSA Camp was changed to Camp Bonifas.

1976년 8월 18일 유엔군사령부측에서는 10명의 경비병을 동반한 5명의 민간인 작업자를 이 지역에 보내서 두 초소 사이의 시계를 양호하게 하기 위하여 매년 있어 왔던 미루나무 가지치기 작업을 지시했습니다. 이들이 도착하자마자, 수적으로 우세한 28명의 북한측 경비병들과 조우하였습니다. 작업은 계획대로 진행되었으나, 약 15분이 경과되자, 그 북한측 경비병들이 유엔사측 요원들을 공격해왔습니다. 약 4분간의 충돌 동안, 공동경비구역 중대장 아더 보니파스 대위와 제1소대장 마크 배렛 중위가 도끼로 잔인하게 살해되었습니다. 당시 희생되었던 보니파스 대위를 추모하기 위하여 저희 기지의 명칭이 캠프 보니파스로 개칭되었습니다.

This is a schematic of the JSA today. This MAC building is the most important building on Conference Row, for it was there that the four essential types of meetings for the supervision of the Armistice were held, and it is currently the site of the General-level officers meetings. I would like you to note the location of these nK guards standing outside the building. They often come to view our tours and it is likely that they will do so today also. We ask you to refrain from making any hand or arm gestures, such as pointing, or attempting verbal communications with these nK soldiers.

현재 판문점 공동경비구역의 개략적인 모습입니다. 군사정전위원회 건물은 회담 장 구역에서 가장 중요한 건물이며, 정전협정 감시에 필수적인 네 가지 종류의 군사회담이 열리는 곳이기도 하고, 현재는 장성급 회담 시 사용하고 있는 건물입니다. 건물밖에 서있는 북한측 경비병들을 주목해 주시기 바랍니다. 그들은 종종 여러분들에게 다가오며, 오늘도 그렇게 할지도 모르겠습니다. 이들에게 손짓을 한다거나 말을 걸려고 하시면 절대 안 되겠습니다.

This is a close-up of two nK Guards. This red collar tab signifies that he is an enlisted, while this yellow collar tab signifies that he is an officer. The "Kyung Moo" armbands these two soldiers are wearing indicate that they are nK Military Police. Under provisions of the Armistice Agreement, all armed personnel within the DMZ must display this form of identification.

2명의 북한측 경비병 모습입니다. 붉은색 계급장을 달고 있는 사람은 북한측 사병이고 노란색 계급장을 달고 있는 사람은 북한측 장교입니다. 경무라고 쓰여진 완장을 착용한 2명의 경비병은 그들이 북한측 군사경찰임을 나타내는 것입니다. 정전협정에 의거 비무장지대 내에서 무장을 하고 있는 인원들은 모두 이와 같은 표시로 자신이 무장하고 있음을 나타내야 합니다.

The UNCSB also provides security and assistance for all visitors on tour of Panmunjom. Our unit reacts quickly to any threats in our area of operations, and performs live combat reconnaissance patrols within our areas of responsibility. Because of the fact that we are the most forwardly deployed unit on the Korean Peninsula, and that we stand face to face with the nK Guards on a daily basis, ROK and the UNC have given us our motto, "IN FRONT OF THEM ALL" which we so proudly state.

유엔사 경비대대는 또한 판문점의 모든 방문객들에 대한 경계 및 지원을 제공해주고 있습니다. 우리 부대는 작전구역 내에서 군사적 위협행위가 있을 경우 즉시 출동하며, 수색 정찰 임무도 수행하고 있습니다. 우리는 한반도에서 최전방에 배치되었기 때문에, 그리고 우리가 매일같이 북한측 경비병들과 얼굴을 마주 보며 근무하고 있기 때문에, 대한민국 정부와 유엔군사령부측에서는 "귀관들은 최전방에서 근무하고 있다"는 자랑스러운 구호를 주었습니다.

Unit 7. Dora Observation Post

도라 전망대(Oral Terrain Briefing)

Key Words

frontal area	전방지역	disarmament	무장해제
observation post(OP)	관측소	front line	전방 전선
southern boundary	남방한계선	unpaved road	비포장 도로
DMZ : Demilitarized Zone	비무장 지대	chimney smoke	굴뚝 연기
		freedom village	자유마을
military demarcation line	군사 분계선	inhabited village	사람이 거주하는 마을
		comprise	구성하다. 포함되다
southern boundary	남방한계선	exempt	면제된, 면제하다
propaganda slogan	선전문구	Joint Security Area	공동경비구역
fence	울타리	binocular	쌍안경

Briefing Text[11)]

Good morning, Ladies and Gentleman. I am First Lieutenant Lee. It's nice to see all of you. From now on, I am going to brief you on the situation of the frontal area.

11) 연합사 『Practical Military English Ⅱ』(비매품), 조용만 · 최병욱 『Military English』. pp.315~317을 참조하여 재구성한 것임.

여러분 안녕하십니까? 이 중위입니다. 여러분을 뵙게 되어서 반갑습니다. 지금부터 전방지역 상황을 보고 드리겠습니다.

You are now at the Dora Observation Post, which is on the Southern Boundary of DMZ. Our 77th Division opens to civilians the front area, Myolgong Museum, the 3rd invasion tunnel including this OP to help them understand the military tension between two Koreas. Seoul is located 43km to the south, and Kaesung, one of the North Korea(NK) cities, located 8km to the North.

여러분들은 지금 도라 관측소에 계시며, 이 관측소는 비무장지대 남방 한계선 상에 위치하고 있습니다. 저희 77사단은 남북한의 군사적 긴장상태를 이해시킬 목적으로 민간인들에게 이 관측소를 포함하여 전방지역, 멸공관 및 제3땅굴을 공개하고 있습니다. 서울은 남쪽으로 43km 떨어져 있으며, 북한의 도시 중 하나인 개성은 북쪽으로 8km 떨어져 있습니다.

First of all, I'd like to explain the Military Demarcation Line. Please give your attention to your front. You can see the small hill with the NK propaganda slogan. The MDL passes the base of the hill, which divides our peninsula, the South and NK. There is no fence for the MDL; yellow signposts just stand every 200meters.

먼저 군사분계선에 대해서 설명 드리겠습니다. 여러분의 전방을 주목해 주시기 바랍니다. 북한 선전 문구가 새겨진 작은 언덕을 보실 수가 있을 것입니다. 우리 한반도를 남한과 북한으로 나누고 있는 군사분계선은 그 언덕 밑을 지나갑니다. 울타리가 없고, 단지 황색 표지판들이 200m 간격으로 설치되어 있습니다.

The origin of the MDL was divided by US and Russian forces at the 38th parallel after World War Ⅱ. It was set to facilitate the Japanese disarmament. It has been almost 00 years since the Korean War and we are the only remaining divided country in the world as the result of Cold War. Although Cold War between the East and the West does not exist any more, NK never gives up its intention to unify Korea by force.

원래 군사분계선은 제2차 세계대전 후에 북위 38도선으로서 미군과 러시아군이 일본군의 무장해제를 위해 설치한 선입니다. 한국전쟁이 일어 난지도 거의 00년이 지났고 현재 우리나라는 냉전의 결과로 분단된 유일한 국가입니다. 동서 진영 간의 냉전은 더 이상 존재하지 않고 있지만 북한은 아직도 무력에 의한 한반도 적화야욕을 버리지 않고 있습니다.

About 35 km to the front from here is the Southern Boundary Line. That is our front line. Please look to your right side. You can see the wire fence along the unpaved road. To the direction of 2 o'clock you can see the NK red flag, or "In Gong Ki" over a high tower. The flag is 15 meters in width and 30 meters in length. The NK propaganda village, or Gi Jung Dong is located there.

이곳으로부터 전방 약 35km 지점이 남방한계선입니다. 그곳이 우리의 최전방 전선입니다. 우측방향을 주목해 주시기 바랍니다. 여러분은 비포장도로를 연하여 설치된 철조망을 보실 수가 있습니다. 2시 방향을 보시면 여러분들은 북한기, 즉 "인공기"를 보실 수가 있습니다. 그 깃발은 폭이 15m이며, 길이가 30m입니다. 북한의 선전마을인 기정동이 그곳에 위치하고 있습니다.

None of the buildings have windows. All the lights are of the same color and are switched on and off at the same time. Chimney smokes and laundries are not visible and they built a number of tall buildings just for us to see. Only soldiers and keepers of the village live there to protect the village.

건물들은 창문이 없습니다. 모든 불빛은 동일한 색깔이며 동시에 불이 들어오고 나갑니다. 어떠한 굴뚝 연기나 세탁물들이 보이지 않습니다. 많은 건물들이 우리들에게 보여 주기 위해서 세운 것입니다. 단지 군인들과 관리인들만이 이 마을을 지키기 위해 거주합니다.

Please look at about 3 km to the right side of it. There is our freedom village, Dae Sung Dong, which is only inhabited village within the DMZ. You can see another high tower with the South Korean national flag, "Tae Keuk Ki" on the top of it. About 00 people comprising about 00 families live there and they are exempt from military services and taxes.

오른쪽으로 약 3km 지점을 보십시오. 우리의 자유의 마을이라고 부르는 대성동이며 비무장지대 내에 주민이 거주하는 유일한 마을입니다. 여러분들은 또 다른 높은 탑과 "태극기"를 보실 수가 있습니다. 약 00세대로 구성된 00명의 주민이 그곳에 거주하며 그들은 국방 및 납세 의무가 면제됩니다.

Look at about 2km to the right side of the flag. You can see a white building in the forest. That is 'the Peace Hall' of Panmujom, otherwise known as the Joint Security Area.

오른쪽 2km 지점을 주목해 주십시오. 여러분들은 숲 속에 흰 건물을 보실 수가 있습니다. 그곳이 공동경비구역이라고 부르는 판문점 내의 '평화의 집'입니다.

To the left side, there is an old rusted locomotive in the forest that ran between NK and South Korea before the Korean War. But now it stands there as the symbol of people's desire for peaceful unification. It has been almost 00 years since the Korean War.

왼쪽을 보시면, 한국전쟁 이전에 남한과 북한을 운행했던 숲 속에 있는 기관차를 보실 수가 있습니다. 그러나 지금은 평화통일에 대한 한국인의 염원의 상징으로 서 있습니다. 한국전쟁이 발발한지 거의 00년이 지났습니다.

This concludes my briefing. Thank you. Now you are free to look around front area through these binoculars on both sides.

이상 보고를 마치겠습니다. 이제 양측에 설치되어 있는 쌍안경을 이용하여 전방지역을 살펴보시기 바랍니다. 감사합니다.

Unit 8. Interview Samples

인터뷰 예문

1. How does the North attempt to create social disorder in the South?
2. What is the North Korea's military strategy toward the South?
3. What is the function of a staff officer?
4. How does a staff officer go about making estimates?
5. What is on SOP and why is it necessary?
6. What is firepower?
7. What is an operational plan and what purpose does it serve?

Key Words

disorder	혼란, 난동	fire support plan	화력지원계획
instigate	선동하다, 부추기다	artificial obstacle	인공장애물
short-term blitzkrieg	단기 속결전	sight	가늠자
self-propelled artillery	자주포병	battlefield surveillance	전장감시
courses of action	방책	OPLAN	작전계획
standing operating procedure (SOP)	예규, 표준임무수행절차	subordinate unit	예하부대
maneuver	기동	company maneuver plan	중대 기동계획

Interview Text[12)]

(1) How does the North attempt to create social disorder in the South?

North Korea uses many methods to create disorder within the south. Its main purpose is to encourage the anti-government mood among the civilian population and military personnel. This is done in part through the distribution of propaganda. The North also attempts to instigate the anti-government feelings by infiltrating into student groups and unions. In addition, the North frequently and persistently demands the withdrawal of the USFK as a condition of U.S.- North Korea negotiations.

북한은 한국 내에서 사회 혼란을 야기하기 위해 어떤 시도를 하는가?

북한은 한국 내에서의 사회 혼란을 야기하기 위해 많은 방법을 사용한다. 주된 목적은 민간인, 군인들 내에 반정부 감정을 악화시키는 것이다. 이것은 선전을 통하여 부분적으로 행해진다. 북한은 또한 학생 단체와 노동단체에 침투하여 반체제 감정을 선동하려고 시도할 것이다. 게다가 북한은 집요하게 북미 협상 조건으로서 주한미군 철수를 요구하고 있다.

(2) What is the North Korea's military strategy toward the South?

The North will conduct a short-term blitzkrieg. The purpose of this type attack is to create panic in the south in the early stages of war by launching several attacks at the same time in forward and rear areas. It will go deep into the South with maneuver forces armed with tanks, armored vehicles and self-propelled artillery. This will give the North the initiative and thus allow them to sweep the entire South before the U.S. reinforcements can arrive.

12) 연합사 『Practical Military English Ⅱ』(비매품)을 참조하여 재구성한 것임.

한국에 대한 북한의 군사전략은 무엇인가?

북한은 단기 속결전을 수행할 것이다. 이 공격 형태의 목적은 전후방 여러 지역을 동시에 공격함으로써 전쟁 초기 단계에서 한국 내에 공황을 야기 시키는 것이다. 전차, 장갑차, 그리고 자주포병으로 무장된 기동력으로 남한을 종심 깊이 공격할 것이다. 이 전략으로 북한은 주도권을 잡고 미증원군이 도착하기 전에 한국 전 지역을 석권하려고 할 것이다.

(3) What is the function of a staff officer?

The basic function of a staff officer is to assist the commander. This is done by making and executing timely decisions, identifying and simplifying complex procedures for the ultimate purpose of accomplishing the mission.

This is done by providing information, making estimates and recommen-dations, preparing plans and issuance of orders, and supervising the implementation of decisions.

참모장교의 역할(기능)은 무엇인가?

기본 역할은 지휘권을 보좌하는 것이다. 이것은 적시적절한 의사결정과 임무수행의 궁극적인 목적을 위하여 복잡한 절차를 단순, 명료화함으로서 수행된다. 이것은 첩보제공, 참모판단 및 건의, 계획 준비와 명령하달, 그리고 감독을 통해서 수행되어야 한다.

(4) How does a staff officer go about making estimates?

It is a part of a staff officer's role to make estimates so that he or she an assist the commander in reaching decisions. Estimates are used to analyze he influence of various factors with the staff officer's field of interest. The officer must identify factors which affect the formulation, analysis and comparison of courses of action. The staff estimates will result in conclusions and recommendations that will identify feasible courses of action.

참모장교는 어떻게 판단하여야 하는가?

판단은 참모장교의 역할의 일부분으로 참모장교가 지휘관의 결심을 도울 수 있게 한다. 판단은 참모장교의 해당 분야에서 다양한 요소에 영향을 주는 것을 분석하는데 활용된다. 참모장교는 방책 구상과 비교 분석 과정에 영향을 주는 요소들을 식별해야만 한다. 참모판단은 실현 가능한 방책을 식별할 수 있는 결론과 건의로 귀결된다.

(5) What is on SOP and why is it necessary?

SOP stands for standing operating procedure. This is a standardized procedure and contains instructions concerning routine or recurring activities. SOP is necessary because it promotes teamwork, efficiency and reduces the need for repetitive instructions. All members of a unit must practice the SOP to train the event of an emergency such as a counter -infiltration operation, or war-time operation.

SOP란 무엇이며 이것이 왜 필요한가?

SOP란 표준임무수행 절차이다. 이것은 표준화된 절차이며, 일상적이거나 반복적인 활동을 포함한다. SOP는 팀웍과 능률을 증진시키고, 반복된 지시를 줄이기 위하여 필요한 것이다. 부대의 모든 구성원은 대침투작전 또는 전시 작전과 같은 위급상황의 SOP를 연습해야만 한다.

(6) What is firepower?

The Commander's concept of operation covers two are as: the scheme of maneuvers and a fire support plan. Fire power is a way to assist in forward movement in an attack as well as a means to reinforce defending capabilities which consist of natural and artificial obstacles. Fire power consists of direct fire and indirect fire. Direct fire includes individual weapons, machine guns and anti-tank weapons where the shooter directly uses a sight and aims the weapon. Indirect fire includes the 105mm and 155mm Howitzer, for example, where the installation and operation of the weapon is under the direction of observation posts or through the use of battlefield surveillance equipment. The success of attacking or defending forces greatly depends upon effective fire support.

화력이란 무엇인가?

지휘관의 작전개념은 두 가지로 볼 수 있는데 그것은 기동력과 화력지원 계획이다. 화력은 공격 기동시 전방을 지원할 뿐만 아니라 (방어시에는) 자연 및 인공 장애물로 구성된 방어 능력을 보강하기 위한 수단이다. 화력은 직접화력과 간접화력으로 구성되어 있다. 직접화력은 개인화기, 기관총, 대전차무기를 포함하며 사수가 직접 가늠자를 사용하여 화기를 조준한다. 간접화력은 105미리, 155미리 곡사포를 포함한다. 예를 들면 화포의 설치 및 운용은 관측소의 지시 또는 전장감시 장비의 사용을 통해서 운용된다. 기동과 방어 부대의 성공은 전적으로 효과적인 화력지원에 달려 있다.

(7) What is an operational plan and what purpose does it serve?

OPLAN includes maneuver plan and firepower plan. The purpose of a maneuver plan is to efficiently execute the deployment, maneuver and employment of the subordinate units in order to achieve the mission. The purpose of a fire power plan is to integrate the company's organization, and support the attached weapon employment plans with the company maneuver plan within the battalion firepower plan.

작전계획은 무엇이며 이것을 수행하는 목적은 무엇인가?

작전계획은 기동계획과 화력계획을 포함한다. 기동계획의 목적은 임무를 완수하기 위하여 병력의 전개와 기동, 예하부대의 운용을 효과적으로 수행하는 것이다. 화력계획의 목적은 중대 조직의 통합과 대대 화력계획 범위 내에서 중대 기동계획에 따른 배속화력 운용계획을 지원하는 것이다.

Unit 9. General Affairs

주요 업무표현

1. The Korean War
2. South Korea- U.S. alliance
3. North Korean nuclear
4. ROK-US exercise
5. Military base management
6. Accidents

Key Words

victim	피해자, 희생자	asymmetric capability	비대칭능력
devoted	바치다, 기울이다	long range missile	장거리 미사일
military alliance	군사적 동맹	PDE(Planning, Decision, Excution) cycle	계획, 결심, 시행 주기
veteran	참전용사	security inspection	보안점검
ROK-US MDT(Mutual Defense Treaty)	한미상호방위조약	FOUO(for official use only)	대외비
provocation	도발	sexually harassing	성추행
combined defense	연합방위	human resource management	인적자원관리
ROK-US SCM	한미연례안보회의	tax evasion	탈세
ROK-US MCM	한미군사위원회	Promotion Board	진급선발위원회
WMD(Weapons of Mass Destruction)	대량살상무기	Islamic fundamentalism	이슬람 근본주의자

Reading Text

(1) The Korean War

We are here today to pay tribute to the victims who devoted themselves defending the nation during Korean war. Someone says the Korean war is Forgotten war but I don't take the freedom for granted at a single moment.

우리 오늘 여기 한국전쟁시 자신을 희생하여 조국을 수호 하셨던 분들을 기리기 위해 모였습니다. 몇몇 사람들은 한국전쟁이 잊힌 전쟁이라고 말하지만 저는 한번도 이 자유를 당연히 여긴 적이 없습니다.

Having spent the last 70 years since the Korean War, the ROK and the U.S. have forged of politics, economy and society, but it is true that the military alliance exists at the center of the relationship for the stability on the Korean Peninsula.

6 · 25전쟁 이후 70여 년이 지난 지금 한미 양국은 정치, 경제, 사회 등 모든 분야에서 폭넓은 협력을 맺고 있지만 그 중심에는 한반도 안정을 위한 군사적 동맹이 있는 것이 사실이다.

South Korea is well recognized for having transformed itself from an impoverished, war-torn society into a developed and democratic nation. The cornerstone of South Korea's global strategy is to embrace international responsibilities and one of the most visible and tangible example has been peacekeeping.

한국은 전쟁으로 인하여 분열된 빈곤한 국가를 발전한 민주주의 국가로 전환한 것으로 잘 알려져 있습니다. 이러한 한국의 세계화 전략에 초석이 되는 것은 국제적 책임을 수용하는 것이며 분명히 실재하는 예가 평화유지 활동입니다.

President Park spent her final day in Colombia offering her thanks to Colombian veterans who fought in the Korean war. Colombia was the only South American country to send troops, and nearly 700 Colombian were either killed or injured in the battle.

박 대통령은 한국전쟁에서 싸웠던 콜롬비아 참전 군인들에게 감사를 표하며, 콜롬 비아 에서의 마지막 날을 보냈습니다. 콜롬비아는 남미국가 중 유일하게 파병한 국가로 약 700여명의 콜롬비아 군인이 전투에서 전사하거나 부상했습니다.

(2) South Korea - United Sates alliance

The ROK-U.S relationship was forged in blood since the Korean War. This year marks the 70th year of the ROK-U.S alliance and it remains steadfast in its commitment to ensuring peace and stability on the Peninsula.

한미관계는 6 · 25이후 혈맹으로 맺어졌습니다. 올해는 한미동맹 70주년을 기념하는 해이며, 한미동맹은 한반도의 평화와 안정의 보장에 크게 기여하며 굳건히 유지되고 있습니다.

SecDef(Secretary of Defence) Carter and Defense minister Han reaffirmed their commitment to the ROK-US MDT(Mutual Defense Treaty)and recognized ROK and US relationship is cemented in blood. They also agreed to adhere to the principle of not rewarding any provocations.

양 장관은 한미상호방위조약에 대한 양국의 공약을 재확인 하였고, 한미관계는 혈맹 이라는 점에 인식을 같이 하였습니다. 또한 어떠한 도발에도 보상하지 않는다는 원칙을 고수하기로 하였습니다.

Secretary of Defense Ashton Carter reaffirmed the United State's resolve and support for the ROK-U.S. alliance and the defense of the Republic of Korea on his first day of visit. The United States is committed to stability in the region and the combined defense of the Republic of Korea. On the peninsula, deterrence and readiness are at a premium.

애쉬톤 카터 미국방부장관은 방문 첫날 한미동맹과 대한민국방어에 대한 미국의 단호한 결심과 지지를 재확인했습니다. 미국은 지역안정과 대한민국의 연합방위를 위해 전념하고 있으며, 한반도의 억제와 전투준비는 최고 상태입니다.

There is concern that the attack on the US ambassador to Seoul could amage the ROK-US alliance witch ensures peace and security on the Korean Peninsula. However, The US envoy made a comment on his twitter that reaffirmed US commitment to the unwavering alliance of the two allies.

마크 리퍼트 피습사건이 한반도의 평화와 안보를 보장하는 한미동맹에 타격을 줄 수 있다는 우려가 제기되고 있습니다. 그러나 리퍼트 대사는 자신의 트위터에 확고한 동맹에 대한 미국의 의지를 재확인 하는 글을 남겼습니다.

Thank you. I'll have the opportunity to meet with President Park during the annual ROK-US SCM slated for this fall. Minister Han and I discussed various ways to contribute to the strengthening of the ROK-US alliance through this MCM.

감사합니다. 올해 가을에 예정되어 있는 연례적인 한미 안보위원회 회의에서 박 대통령님과 대화할 기회가 있을 것으로 예상됩니다. 한 장관님과 저는 이번 군사위원회 회의를 통해 한미동맹 강화에 기여할 수 있는 다양한 방안에 대해 논의하였습니다.

* 한미연례안보회의(SCM): ROK-US annual security consultative meeting.
* 한미군사위원회(MCM): Military Committee Meeting

(3) North Korean nuclear

The rise of China and North Korea's Weapons of Mass Destruction are major issues that need be discussed in detail during today's meeting. We all know very well that North Korea is run under a dictatorship and that is a isolated state.

중국의 부상과 북한의 대량살상무기는 오늘 회의에서 상세하게 다루어져야 하는 사항들입니다. 우리 모두 북한은 독재체제라는 것과 고립된 국가라는 것을 매우 잘 알고 있습니다.

North Korea's growing asymmetric capabilities present challenging and complex threats to the ROK-US Alliance. North Korea maintains the world's largest special forces and it also has one of the largest chemical weapon.

북한의 증가하고 있는 비대칭전력은 한미동맹에 도전적이며 복잡한 위협입니다. 북한은 세계 최대 특수전 병력을 유지하고 있으며 최대 화학무기 재고 보유국 중 하나이기도 합니다.

North Korea's nuclear weapons and long range missiles are threats to South Korea and world. It still denies that it is behind the sinking of Cheonan and not responding to the proposal for the inter-Korean talks.

북한의 핵 개발과 장거리 미사일은 한국과 전 세계에 위협이 되고 있습니다. 북한은 천안함 피격이 자신들의 소행임을 아직까지 부인하고 있으며 남북대화에 응하지 않고 있습니다.

As I understand, the levels of interoperability we achieved with our partners in the Korean Theater during the most recent North Korean missile launch were unprecedented; however, there is more work to be done.

저는 최근 북한의 미사일 발사 당시 한국 전구 내에서 우리의 우방국들과 함께 이룬 상호운용성의 수준은 전례 없는 것이었다고 생각합니다. 그러나 아직 넘어야 할 산들이 많습니다.

North Korea must break a stand off over their nuclear program and be a responsible member of the international society. North Korea will have to join our effort to achieve peace and unification of the Korean peninsula by building trust in the inter Korean relationship.

북한은 핵으로 인한 교착상태를 타개하고 국제사회의 일원이 되어야 합니다. 북한은 남북 간 신뢰를 구축하여 한반도의 평화와 통일을 만들어 가려는 우리의 노력에 동참해야 할 것입니다.

(4) ROK-US exercise

We will compile recommendations from every section for the combined ROK US exercise. Nagative response is required. Keep in mind we should have good coordination with them throughout the duration of the exercise.

이번 한미 연합훈련에 대한 각 부서의 건의사항을 수렴하겠습니다. 해당 사항이 없으면 없다고 보고해 주십시오. 훈련 기간 동안 미국 측과 업무협조를 잘할 것을 명심하십시오.

C4I means Command, Control, Communication, Computer and Intelligence. The basic concept of employment of C4I is as follows. It starts with tuational Awareness, Situational Understanding and Battle space Dominance.

C4I는 지휘, 통제, 통신, 컴퓨터, 정보를 뜻합니다. C4I의 기본활용개념은 다음과 같습니다, 상황인식을 시작으로 상황이해, 전장지배순입니다.

First of all, plan operations based on the PDE cycle, then make decisions based on the collected information and processed intelligence, and finally, carry out operations in accordance with the operational environment. We are currently conducting 24 hour operations in two 12 hour shifts, so report the most recent BUB(Battle Update Briefing) during shift change.

먼저 PDE 주기에 맞춰 작전을 기획하고, 수집된 첩보와 가공된 정보에 따라 결심을 한 뒤 작전환경에 따라 작전을 수행하면 됩니다. 우리는 현재 12시간 근무교대로 주야 연속작전을 시행하고 있으므로 교대시 최신화 된 전투상황 보고를 하시면 됩니다.

Key Resolve is a combined exercise to strengthen the capabilities for prompt deployment of the U.S. Augmentation Forces to the Korean peninsula. Since Team Spirit training was cancelled in 1994, RSOI was initiated and the name has been changed to Key Resolve since 2008.

키리졸브는 한반도에 미군 증원군을 신속하게 전개할 수 있는 능력을 강화하기 위한 연합훈련입니다. 팀스피리트 훈련이 1994년 취소된 이후 이를 대체해 연합전시증원연습이 시작되었으며 2008년부터는 키리졸브라는 명칭을 사용합니다.

(5) Military base management

CPT Lee is quickly adjusting to operations officer position. Although he is learning over the shoulder, he tries to take care of his own work in order to better conduct given tasks.

이 대위는 작전장교 보직에 잘 적응하고 있습니다. 그는 어깨 너머로 일을 배우고는 중이지만, 부여된 임무를 보다 잘 수행하기 위해 자기 일을 스스로 처리하려고 노력합니다.

As a leader CDRs should build teamwork and trust within their units. Once trust is violated, a leader becomes ineffective. Meanwhile, soldiers must trust and have confidence in their leaders.

지휘관은 리더로서 부대 내 팀웍과 신뢰를 구축해야 합니다. 신뢰를 상실하게 되면 리더는 효과적으로 지휘할 수 없게 됩니다. 한편 병사들은 그들의 리더에 대하여 신뢰와 확신을 가져야 합니다.

We are going to have a security inspection in a few day's Make sure that all the classified, confidential above, as well as FOUO(for official use only) are placed in double locked storage boxes. With OPSEC in mind, check all of your USB'S and floppy disks.

며칠 뒤에 보안점검을 하게 됩니다. 3급 이상의 비밀은 물론 대외비도 이중 잠금장치가 된 보관함에 보관해야 합니다. 작전보안에 유념하며 USB와 디스켓이 모두 확인하시기 바랍니다.

At around 2000 last night, two soldiers, while illegally operating a moped off- post, were struck by a car driven by a civilian. Two soldiers were given an alcohol test by the police and failed.

어제 23시경 2명의 병사가 음주 후 부대 밖에서 오토바이를 운전하다가 민간인이 운전하는 차와 충돌하였습니다. 2명의 병사는 경찰에 의해 음주측정을 받았으며 음주운전인 것으로 판명되었습니다.

Master Sergeant Harris may be tried by US Military tribunal of sexually harassing his subordinate. Although he is pleading innocent to the charges for crime, non judicial punishment will be imposed such as a salary reduction, degradation as well as extra duties.

헤리스 상사는 부하직원에 대한 성추행 혐의로 미 군사 특별 재판소에서 재판을 받을 수도 있습니다. 그는 무죄를 주장하고 있지만 감봉은 물론이고 강등, 추가 근무 등 비사법적 처벌 또는 징계가 내려질 겁입니다.

Let me introduce you my replacement who is standing right next to me. Before coming here, Captain Lee worked at the Army Head- quarters from 2020~2022, and was in charge of human resource management. I think he is an excellent new comer with various skills, and I hope all of you help him get adjusted to the new environment.

지금 제 옆에 서있는 제 후임자에 대해 소개드리겠습니다. 여기에 오기 전에 이 대위는 2020년도에서 2022년까지 육군본부에서 근무를 하였으며, 인적자원관리 업무를 했었습니다. 이 대위는 다양한 실력을 겸비한 우수한 전입자이며 여러분들이 이 대위가 새로운 환경에 적응할 수 있게끔 도움을 주었으면 합니다.

Promotion Board is working on officer Evaluation Report. Officers who get marginal personal evaluation will not be in a promotion zone this year. Keep in mind that personnel's ability is more important than seniority.

진급선발위원회에서 인사평정을 작성하고 있습니다. 최저평정을 받는 장교는 올해 진급 대상자가 아닙니다. 능력이 연공서열보다 중요합니다.

(6) Accidents

Recently, there was a car accident in Seoul. So, the police decided to conduct a through probe into the matter. The priority of the police was on finding out whether the driver took all necessary protective measures while driving. To prevent such accidents from happening in the future, one needs to be familiar with the regulations of the city road.

최근, 서울시에서 차량사고가 발생하였습니다. 이에 경찰은 사고에 대해 철저히 조사하기로 하였습니다. 경찰의 우선순위는 운전자가 운전 중에 모든 필요한 예방책을 지켰는지에 대해 초점이 맞추어져 있습니다. 미래에 이러한 사고의 재발을 방지하기 위해, 도시 도로법을 숙지하고 있어야 합니다.

Mr. Kim, who is believed to be a person in the higher bracket, was arrested on Tuesday on charges of tax evasion. The prosecution official in harge of investing the case, said "I don't understand why rich people commit such wrongdoing, when all the poorer people pay their taxes all the time."

고 소득자로 알려진 김씨는 탈세 혐의로 화요일 체포되었습니다. 금번 사건의 조사를 담당하는 검찰관계자는 '왜 부유한 사람들이 이러한 범행을 저지르는지 모르겠습니다. 가난한 사람들은 오히려 꾸준히 세금을 내는데 말이죠'라고 언급하였습니다.

Aircraft 700 has crashed into the Himalayan mountains. Where abouts or the aircraft is unknown. Out of 300 passengers, personnel unaccounted for are 20 and 48 personnel remain to be contacted and for by contacting their family members.

항공기 700이 히말라야 산에 추락하였습니다. 항공기의 위치는 아직 파악이 되지 않았습니다. 300명의 탑승 인원 중 20명은 소재가 확인이 안 된 인원이며 48명은 가족에게 연락하여 파악을 해야 합니다.

The issues of Islamic fundamentalism and suicide bombing are gaining international attention. So, today we will discuss the issue of terrorism in the 21st and what actions we can take to prevent the spread of terrorism.

이슬람 근본주의 및 자살 폭탄 테러의 문제들은 국제적인 관심을 끌고 있습니다. 그래서, 오늘 우리는 21세기의 테러 확산에 대해 우리가 취할 수 있는 어떠한 활동들이 있는지 토의 해보겠습니다.

UNIT 10. Progression of Meeting

회의 진행

Key Words

UFS	을지 자유의 방패	mobilization planning	동원계획
IPC	최초계획 회의	intermission	휴식
Y2K	컴퓨터 인식 버그	representative	대표자
rivercrossing operation	도하작전	share(d) expenses by divisions	
enemy attack scheme	적 공격 기도		처부별 분담금
counterattack plan	역습계획	summary	요약

Reading Text

(1) Welcoming Address

Good morning everyone? Thank you for coming to the CFC Foal Eagle Conference. I would like to announce that the meeting has officially begun.

I am privileged (honored) to have the opportunity to welcom you all to this conference and participate with you in this opening ceremony.

It is a real honor and privilege for us(me) to welcome you.

I would like to welcome all action officers who are participating in UFS'23 IPC.

환영 인사

여러분, 안녕하십니까? 연합사 독수리훈련 회의에 참석하신 데 대해 감사드리며 회의를 시작하도록 하겠습니다.

여러분을 환영할 수 있는 것을 큰 영광으로 생각합니다.

여러분의 개회식 참석을 환영하며 제가 이 행사의 사회를 보게 된 것을 영광으로 생각합니다.

UFS'23 최초계획 회의에 참석한 모든 실무 장교님들을 환영합니다.

(2) Self Introduction

I am Colonel Lee, C-1, and it is a great honor to be the chair person for today's meeting.

I am the Chief of Plans and Operations Division, C-4, and I'm going to be the chair person for today's session.

Let me introduce myself. I am the chairman for this morning's session. I am Colonel Lee from the Operations Division, and my chairman is Colonel Robinson from Plans Division.

I am Colonel Lee from the Operations Division. It is a great pleasure for me to share the chairmanship with Colonel Robinson from the Plans Division

사회자 소개

저는 인사참모부 이 대령입니다. 제가 금일 회의의 사회를 볼 수 있게 된 것을 큰 영광으로 생각합니다.

오늘 회의를 진행할 군수참모부 계획운영처장입니다.

제 소개를 하겠습니다. 저는 오늘 아침 회의를 주관할 작전처의 이 대령이고 저와 같이 주관할 사람은 계획처의 로빈슨 대령입니다.

저는 작전처의 이 대령입니다. 계획처의 로빈슨 대령과 같이 회의를 주관하게 돼서 매우 기쁩니다.

(3) Opening Address/Outline of Conference

The purpose of our meeting today is to plan a picnic.

The aim of this session is to review the present state of knowledge and to discuss related theories.

In view of the shortness of time and our tardiness in getting started, I think we should go right into the papers.

We are gathered today to discuss the problem of Y2K.

The symposium is on "The Improvement in the Ability to Execute a Rivercrossing Operations."

Our program this morning consists of presentations by a specialist from the Defense Intelligence Command who will discuss the enemy's attack scheme.

The purpose of our meeting today is to discuss the division defense plan. Let us start this meeting with a discussion on FEBA A.

It is our pleasure to have this opportunity to discuss some of the problems that we see in the division's counterattack plan.

To begin with, we will spend a few minutes discussing some of the problems regarding the timing of the counterattack.

The plan for this panel discussion is as follows. First, we would like to spend a few minutes discussing the concept of operations for the defense. Second, we will review the possible enemy courses of action. Third, we will look at the terrain analysis. Finally(Lastly), we would like to entertain questions from the floor.

We are in the middle of the panel discussion on the "Effective Mobilization Planning."

Now It is time to begin the panel discussion on "NBC Protection."

개회사/사회의 개요

오늘 회의의 목적은 야유회 계획에 대한 것입니다.

이번 회의의 목적은 현 숙지 상태를 확인하고 관련된 이론에 대해 토의하는 데 있습니다.

시간도 부족하고 늦게 시작했기 때문에 바로 본회의로 들어가겠습니다.

Y2K 문제에 대해 토의하기 위해 오늘 이 자리에 모였습니다.

본 토론회는 도하작전 능력향상에 관한 것입니다.

오늘 오전에는 정보사령부에서 오신 전문가의 적의 공격기도에 대한 발표가 있겠습니다.

오늘 회의의 목적은 사단방어계획에 대해 토론하는 데 있습니다. FEBA "A" 지역에 대한 토론으로 본회의를 하겠습니다.

사단 역공격 계획상의 몇 가지 문제점들에 대해 토론할 수 있는 기회를 갖게 되어 기쁩니다. 우선, 역공격 시기에 관한 문제들에 대해 몇 분 정도 토론하겠습니다.

본 토론회의 예정은 첫째로, 몇 분 정도 방어작전 개념에 대해 토론했으면 좋겠고 둘째로, 적의 가능한 방책을 고려해야 하겠으며 셋째로, 지형 분석 검토순으로 이어지겠습니다. 그리고 마지막으로, 참석자들의 질문을 받겠습니다.

효과적인 동원계획에 대한 토론회 중입니다.

이제 화생방 방호에 대한 토론회를 시작할 시간이 되었습니다.

(4) Introduction of Speaker

It gives me great pleasure to introduce today's guest speaker, Dr. Kim from the Cheongju University. We feel grateful that he has taken time from his busy schedule to spend this hour with us.

We are especially honored to have Lieutenant General Lee of JCS with us today.

May I present to you Major General Kim, our guest speaker.

We have invited Colonel Watson to speak to us about the next exercise.

강사소개

오늘의 초대 강사이신 청주대 김 박사님을 소개할 수 있어 매우 기쁘게 생각합니다. 우리와 같이 이 자리를 하시기 위해 바쁜 일정에도 불구하고 시간을 내주신 것에 대해 감사드립니다.

합동참모본부의 이 장군님을 이 자리에 모실 수 있어 영광입니다.

초대 강사이신 김 장군님을 소개합니다.

다음 연습에 대해 발표해 주실 왓슨 대령님을 초대했습니다.

(5) Progression of Meeting

Now, we'll begin the presentation based on the issues discussed up to this point.

The speakers have been asked to limit their remarks to one minute for the keynote speech and four minutes for each subsequent speech. We hope this will leave 10 minutes at the end of the session for questions from the floor.

The agenda of today's session is shown on the chart. From now on, we will start with the summary report.

Today we are presenting a panel discussion on "Ways to Improve Work Efficiency."

We will discuss ways to improve the operation of the welfare committee.

This slide is meant to give you an outline of the unit's introduction.

I would like to offer(give) a short(brief) description of the concept of operation.

회의진행

자 그럼, 지금까지 토의한 내용에 대해 발표를 시작하겠습니다.

발표자들의 요약 발표시간은 1분, 그리고 이어지는 발표시간 4분이 주어져 있습니다. 그리고 발표가 끝나는 대로 참석자들은 10분 동안 질문할 시간을 갖게 될 것입니다.

오늘 회의 일정은 차트에서 보시는 바와 같습니다. 지금부터 개요 보고 브리핑이 있겠습니다.

오늘은 업무 효율성 향상방안에 대한 토의가 있겠습니다.

복지위원회 운영 개선방안에 대한 토의를 하겠습니다.

이 슬라이드는 부대 소개에 대한 개요를 보여 주고 있습니다.

작전개념에 대해 짧게(간결하게) 말씀드리고자 합니다.

(6) Intermission

I think we need a break. Let's have a recess and meeting in 20 minutes.

Let's have a break and resume the meeting at 14:00.

Unless someone has a pressing question, I think we should pause for a break now.

This session will now recess for a break.

Since we are running 30 minutes behind schedule, we won't have an intermission.

휴식

휴식이 필요한 것 같습니다. 20분간 휴식을 갖고 다시 회의를 시작하겠습니다.

잠시 휴식을 갖고 14시부터 다시 회의를 시작하겠습니다.

급한 질문이 없으면 지금 휴식을 갖겠습니다.

회의 중 휴식을 갖겠습니다.

예정보다 30분 늦었으니 중간 휴식을 생략하겠습니다.

(7) Question/Answer

We ask that your questions be brief and to the point and that you introduce yourself by name, rank, and unit.

Are there any questions or comments?

Would anyone like to comment on A Regiment's paper?

I hope the audience will participate in the discussion by raising their hands.

Are there any questions for Lieutenant Colonel Kang?

If there aren't any questions, let's move on.

I think we are beginning to lose sight of the main point. May I refocus your attention to our subject?

We are running a little behind schedule. Let's have only one more question, please.

Since we are pressed for time, please make your remarks brief.

질의/응답

질문들은 간단하면서도 요점에서 벗어나지 않게 해주시고, 자신을 소개할 때 이름, 계급, 그리고 소속부대를 밝혀 주시기 바랍니다.

질문이나 의견 제시하실 분 있습니까?

A 연대의 자료에 대해 의견을 말하실 분 계십니까?

청중은 거수로 토론에 참여해 주셨으면 합니다.

강 중령님께 질문하실 분 계십니까? 없으면 계속 진행하겠습니다.

죄송합니다만 주제를 벗어나고 있는 것 같습니다. 우리가 토의하고 있는 주제를 재인식해 주시기 바랍니다.

주어진 시간이 얼마 남지 않았습니다. 질문 하나만 더 받도록 하겠습니다.

(8) Result of Discussion

The representative of each group has to give their presentation on the result of the discussion for ten minutes.

We have reached the end of our time limit. It's already five o'clock. We need to come to a conclusion.

I would like to conclude with a few general remarks.

Let me conclude by making an observation concerning the main areas of interest for future policy decisions mentioned by the participants.

May I conclude my speech with a few more remarks on the contents of the presentation given by the speakers?

In conclusion of my discussion today, I'd like to emphasize again the importance of cooperation.

토의결과

각 부처의 대표들께서는 토론 결과에 대하여 10분 간의 발표를 해주시기 바랍니다.

이제 끝낼 시간이 되었습니다. 벌써 5시가 되었군요. 이제 회의의 결론을 내려야겠습니다.

몇 가지 일반적인 의견을 제시하면서 결론짓겠습니다.

참석자들께서 언급하신 차후 정책결정 시 주요 관심 분야에 대한 소견을 제시하면서 결론짓겠습니다.

발표자들의 발표 내용에 대한 의견을 제시하면서 결론지어도 되겠습니까?

협조의 중요성을 다시 한 번 강조하면서 오늘의 토론을 결론짓겠습니다.

(9) Vote

We will now proceed to vote on the amending the standard for evaluating promotions. All in favor of the amendment will say “Aye” by order. Those opposed will say “Nay.”

The vote is on the motion to(adoption of) share(d) expenses by divisions. Those in favor of the motion(adopting) will rise.

Those opposed remain seated.

The ayes have it and the motion is carried (adopted).

The nays have it and the motion is lost.

There are 32 in the affirmative(negative) and 30 in the negative (affirmative). The affirmative(negative) has it and the motion is carried(lost)

투표

진급 심사기준 수정안에 대한 표결이 있겠습니다. 수정안에 동의하시는 분들은 순서대로 "예"라고 하시고 반대하시는 분들은 "아니오"라고 대답해 주십시오.

처부별 분담금에 대한 안건의 채택 여부입니다. 안건 채택에 찬성하시는 분은 일어서 주시고 반대하시는 분은 앉아 계십시오.

찬성하는 분들이 더 많으므로 안건을 가결하겠습니다.

반대하는 분들이 더 많으므로 안건을 부결하겠습니다.

32명이 찬성(반대)했고 30명이 반대(찬성) 했습니다. 찬성(반대)하는 분들이 더 많으므로 안건을 가결(부결)하겠습니다.

(10) Closing Address

I would like to close the meeting by thanking all the speakers and attendees who have taken part in this meeting, especially Major Kim, today's briefer. Again, thank you very much.

To fully summarize this conference is an impossible task. So many valuable contributions cannot be mentioned in just three or four minutes.

We have come to the end of the conference. I'll give you a meaningful summary of it.

We have had an excellent symposium. It is quite impossible for me to summarize the proceedings, but I would like to mention some of the principal points that have emerged from some of the papers.

I am very pleased to have been exposed to such a wide variety of viewpoints during this conference.

It is time for us to close this session.

폐회사

이상으로 회의를 마치면서 이번 회의에 참석해 주신 발표자와 참석자 여러분께 감사의 말씀을 드리고, 특히 오늘 브리핑을 해주신 김 소령에게 감사드립니다. 다시 한 번 감사드립니다.

본 회의를 전부 요약해서 말하기란 불가능합니다. 그 많은 유익한 내용들을 불과 3~4분 만에 언급할 수는 없을 것입니다.

회의를 마칠 때가 왔습니다. 제가 회의를 의미 있게 요약하겠습니다.

훌륭한 회의였습니다. 회의록을 요약하기란 불가능하겠지만, 자료에서 도출된 몇 가지 주요 상황을 말하고자 합니다.

본회의가 진행되는 동안 여러 가지 다양한 관점을 접할 수 있어서 좋았습니다.

자 이제 끝내도록 합시다.

UNIT 11. UN Introduction

유엔 소개

Key Words

legitimate government	합법정부	gravestone	묘비
Security Council	안전보장이사회	memorial tower	위령탑
ethnicity	민족	cemetery	묘지
hot pepper	매운 고추장	Unknown Soldier	무명용사
fluctuated	변동을 거듭하다	commemorative event	기념행사
United Nations Memorial Cemetery in Korea (UNMCK)	재한 유엔기념공원	crematorium facility	화장장 시설

Reading Text[13)]

United Nations(United Nations)

The United Nations is organization with authority established in 1946. The most powerful collective security mechanism developed by mankind since World War II. In 1950, the military was dispatched to repel the illegal invasion of the South by North Korea. The United Nations Command is still located in Seoul.

13) 황인희(2023). 『감사합니다. 잊지 않겠습니다』와 박휘락(2021). 『국가안보론』을 참조하여 재구성한 것임.

국제연합(유엔)

유엔은 1946년 창설된 권위 있는 조직이다. 제2차 세계대전 이후 현재까지 인류가 개발한 가장 강력한 집단안보기구이다. 1950년 북한의 불법남침을 격퇴하기 위하여 군대까지 파견하였다. 지금도 그 유엔군사령부가 서울에 있다.

Background of establishment

After experiencing World War II, which resulted in an unprecedented disaster in which nuclear weapons were used, the United Nations It was established under the agreement of the international community that such a calamity cannot be repeated any more all. In particular, it was developed in the direction of strengthening sanctions against countries that harm the stability and peace of the international community. The League of Nations was established in 1920, right after World War I, and there was no means to act if member states withdrew. For example, when Japan invaded Manchuria in 1933, 40 out of 42 countries asked Japan to withdraw, but there was no way to do it when Japan withdrew from the League of Nations.

설립배경

유엔은 핵무기까지 사용된 전대미문의 재앙을 초래한 제2차 세계대전을 겪고 나서 이와 같은 참화를 더 이상 반복할 수 없다는 국제사회의 합의에 의하여 설립되었다. 특히 국제사회의 안정과 평화를 해치는 국가에 대한 제재를 강화하는 방향으로 발전시켰다. 국제연맹은 제1차 세계대전 직후인 1920년 설립되었는데, 회원국이 탈퇴해 버리면 조치할 수 있는 수단이 없었다. 예를 들면, 1933년 일본이 만주를 침략하자 42개국 중 40개국이 일본에게 철수할 것을 촉구하였지만 일본이 국제연맹을 탈퇴해 버리자 어떻게 할 수 있는 방법이 없었다.

The League of Nations imposed economic sanctions to prevent Italy's invasion of Ethiopia, but in the meantime, Italy conquered Egypt and there was no way to restore it. Under pressure from the international community, Italy also withdrew from the League of Nations. And, within the framework of the League of Nations, Britain and France tried to guarantee peace by promoting negotiations with Adolf Hitler and ceding the Czech Republic, but in the end, World War II broke out by invading Poland in 1939.

국제연맹은 이탈리아의 이디오피아 침공을 예방하기 위하여 경제제재를 부과하여 하였으나 그 사이에 이탈리아가 이집트를 정복해버리자 원상회복시킬 방법이 없었다. 국제사회에서 압력을 가하자 이탈리아도 국제연맹을 탈퇴해버렸다. 그리고 국제연맹의 틀 속에서 영국과 프랑스가 히틀러 (Adolf Hitler)와의 협상을 추진하여 체코를 할양하면서까지 평화를 보장하고자 했으나, 결국 1939년 폴란드를 침공함으로써 제2차 세계대전이 발발하였다.

Recognizing the need for a global collective security system despite the failure of the League of Nations, the US State Department began preparing drafts for it in 1939. The outline of its contents was released under the title “Declaration by United Nations” in 1941, when the war was in full swing. This was later called 'The Atlantic Charter'.

The key content of this declaration is the expansion of the territory to not admit any change in the territory against the will of the people or to restore the people's government. and to disarm the aggressor.

On this basis, the United Nations Declaration was signed on January 1, 1942 by the representatives of the United States, Great Britain, the Soviet Union, and China, which were waging war against Germany and Japan. The next day, 22 additional countries signed it, bringing 26 countries to join, setting the basic direction for the establishment of the United Nations. The core content of the Atlantic Declaration was the concept of implementing collective security with the four countries of the United States, Britain, the Soviet Union, and China as the 'Four Policemen'.

국제연맹의 실패에도 불구하고 세계적인 집단안보체제가 필요하다는 인식하에 1939년 미 국무성에서 그에 관한 초안을 마련하기 시작하였다. 전쟁이 한창이던 1941년 "국제연합의 선언(Declaration by United Nations)"이라는 제목으로 그 내용의 윤곽이 공개되었다. 이것은 나중에 '대서양 헌장(The Atlantic Charter)'으로 불렸다.

이 선언의 핵심내용은 영토의 확장이나 주민의 의사에 반하는 영토의 변경을 인정하지 않고, 주민들의 정부를 복원하며, 침략국을 무장해제한다는 등의 내용이었다.

이에 근거하여 독일과 일본에 대항하여 전쟁을 수행하던 미국, 영국, 소련, 중국의 대표가 1942년 1월 1일 '국제연합 선언'에 서명했다. 다음날 22개국이 추가로 서명함으로써 26개국이 동참하게 되어 유엔의 설립을 위한 기본적인 방향이 설정되었다. 대서양 선언의 핵심적인 내용은 '4명의 경찰(Four Policemen)'로서, 미국, 영국, 소련, 중국의 4개 국가가 중심이 되어 집단안보를 구현한다는 개념이었다.

Then, between September 21 and October 7, 1944, the four powers negotiated the basic matters of the goals, structure and functioning of the United Nations.

On April 25, 1945, a meeting on the establishment of the United Nations was held in San Francisco in the presence of 50 countries, and a draft of the United Nations Charter was agreed upon. Then, on October 24, 1945, the five permanent members of the Security Council (France was added later) and the remaining 46 signed the UN formally. The first General Assembly was held in London on 10 January 1946.

그리고 나서 1944년 9월 21일~10월 7일 사이에 4개 강대국들은 국제연합의 목표, 구조, 기능에 관한 기본적인 사항을 협의하였다.

1945년 4월 25일 샌프란시스코에서 50개국이 참가한 가운에 유엔의 설립에 관한 회의를 실시하여 유엔헌장 초안에 합의하였다. 그런 다음에 안전보장이사회 5대 상임이사국(프랑스는 나중에 추가)과 나머지 46개국이 서명하여 1945년 10월 24일 유엔이 공식적으로 출범하였다. 1946년 1월 10일 런던에서 첫 총회를 개최하였다

Organized the first UN Forces since the founding of the United Nations

In May 1948, the United Nations formed the United Nations Temporary Commission on Korea' to hold elections under UN supervision, and then recognized the Republic of Korea as 'the only legitimate government on the Korean Peninsula'.

창설이후 최초로 연합군 조직

유엔은 1948년 5월 '유엔한국임시위원단'을 구성해 유엔 감시 아래 선거를 치르게 한 후 대한민국을 '한반도에서 유일한 합법 정부'로 승인했습니다.

On June 25, 1950, when North Korea invaded the Republic of Korea, the very next day the United Nations held an emergency Security Council (hereinafter referred to as the Security Council). The Security Council has called on the North Korean military to "immediately stop hostilities and withdraw to the north of the 38th parallel." But North Korea ignored that recommendation. The Security Council met again on the 28th. On that occasion, they resolved to "take joint action to ensure world peace and freedom on the Korean Peninsula." This resolution meant that the United Nations would form the first international coalition to defeat a common enemy. The goal of the June 28 UN resolution was to restore peace by imposing military sanctions on acts that destroy world peace.

1950년 6월 25일 북한이 대한민국을 침략하자 바로 다음 날 유엔은 긴급 안전보장이사회(이하 안보리)를 개최했습니다. 안보리는 북한군에게 "적대 행위를 즉각 중지할 것과 38도선 이북으로 철수할 것"을 요구하였습니다. 하지만 북한은 그 권고를 무시하였지요. 안보리는 28일에 다시 모였습니다. 그 자리에서 "세계 평화와 한반도의 자유를 보장하기 위해 공동 행동"하기로 결의하였습니다. 이 결의는 유엔이 최초로 국제적으로 연합군을 조직하여 공동의 적을 무찌르겠다는 뜻이었습니다. 6월 28일 유엔 결의의 목표는 세계 평화를 파괴하는 행위에 대해 군사적 제재를 가함으로써 평화를 회복하는 것이었습니다.

With the resolution of June 28, the Korean War expanded from a "war between North Korea and the Republic of Korea" to a "war between North Korea and the United Nations". The United Nations created the 'United Nations Forces Command' and appointed MacArthur as its commander. In response to the UN resolution, 16 countries sent troops to South Korea, and six countries sent medical staff.

이 결의로써 6·25전쟁은 '북한과 대한민국의 전쟁'에서 북한 대 유엔의 전쟁으로 확대되었습니다. 유엔은 유엔군사령부를 만들고 맥아더를 사령관으로 임명하였습니다. 유엔의 결의에 호응한 자유 진영 열여섯 개 나라가 한국으로 군대를 보냈고 여섯 나라가 의료진을 파견했어요.

General MacArthur, the commander of the UN forces, believed that in order for a participating nation to make a real contribution to the war, its troops would have to be 'about 1,000 men capable of adequately fighting, or one battalion'. In addition, it was decided that the participating countries should 'arrive to the area of operation with 60 days' worth of supplies fully armed' and 'the home country should be responsible for all logistical support'.

유엔군 사령관 맥아더 장군은, 참전하는 국가가 실질적으로 전쟁에 기여하기 위해서는 병력이 '적절히 전투를 치를 수 있는 약 1,000명, 즉 1개 보병 대대는 되어야 한다고 생각했습니다. 또 참전국은 완전히 무장한 상태에서 60일분의 보급품을 가지고 작전 지역에 도착해야 하며, 본국은 모든 병참 지원에 책임을 질 수 있어야 한다'라고 판단했습니다.

Among the UN member states other than the United States, the first countries to pledge military aid were the Commonwealth countries. Taiwan also said it would send ground troops. However, at that time, China's civil war had just ended, and the United States refused Taiwan's participation in the war due to political problems and the weakening of Taiwan's fighting power.

미국 이외의 유엔 회원국 중 가장 먼저 군사 지원을 약속한 나라는 영연방 국가들이었습니다. 타이완도 지상군을 파견하겠다고 했지요. 그러나 그때는 중국의 국공내전이 끝난 지 얼마 되지 않은 때라 미국은 정치적인 문제와 타이완의 전투력 약화 등을 이유로 타이완의 참전을 거부했습니다.

For the first time since the establishment of the United Nations, twenty-two countries have gathered under the flag of the United Nations. They fought to protect freedom and peace. Although the purpose of participating in the war was the same, each country had great differences not only in language and ethnicity, but also in customs and traditions. There were many difficulties in distributing them because of their religious differences as well as their eating habits. Turkish troops, mostly

Muslims, did not eat pork, and Indian troops who came to provide Medical Support made it a taboo to eat beef. The French, Belgian and Dutch forces wanted more bread and potatoes than the Americans, and the Thais had to provide rice and hot pepper paste.

유엔 창설 이후 처음으로 스물두 개 나라가 유엔기 아래 모였습니다. 그들은 자유와 평화를 지키기 위해 참전한 것입니다. 이렇게 참전 목적은 같았지만 각 나라는 언어와 민족뿐만 아니라 풍속과 전통에는 큰 차이가 있었습니다. 종교적 차이는 물론 식성마저 달라 보급에도 많은 어려움이 있었지요. 대부분 무슬림인 튀르키예군은 돼지고기를, 힌두교도가 많은 인도군은 쇠고기를 안 먹었습니다. 프랑스 · 벨기에 · 네덜란드군은 미군보다 더 많은 빵과 감자를 원했고 태국군에게는 쌀과 매운 고추장을 제공해야 했습니다.

The Korean War was not just a war for the people living on the Korean Peninsula or for the soldiers who participated in the war. It was an international war in which twenty-two friendly countries and two enemy countries fought intertwined on the Korean Peninsula. The large sea outside our territorial waters was also bustling with transporting numerous troops and supplies, and warships coming and going for shifts. Also, there was no empty space on the sky road to transport the wounded to the hospital and to carry military supplies. Not only that, as charter prices fluctuated, the whole world had to listen to what is happening on the Korean Peninsula every day.

6 · 25전쟁은 한반도에 사는 사람들이나 참전국 장병들만의 전쟁이 아니었습니다. 아군 스물 두 나라, 적군 두 나라가 한반도에서 얽혀 싸운 국제 전쟁이었어요. 우리 영해 밖 큰 바다도 수많은 병력과 군수물자를 실어나르느라, 군함이 교대를 위해 오고 가느라 부산했습니다. 또 하늘길도 부상자를 병원으로 후송하느라 군수물자 나르

느라 비어 있을 틈이 없었지요. 뿐만 아니라 전세가 엎치락뒤치락하는 바람에 전 세계가 한반도에서 일어나는 사건에 날마다 귀를 기울여야 했습니다.

The Korean War was a tremendous, global event that affected their families, people around them, and countless others, whether veterans were killed or survived.

6·25전쟁은 참전 장병이 전사했든 살아남았든 그 가족과 주변 사람들, 그 수많은 사람에게 영향을 준 엄청난 세계적인 사건이었습니다.

United Nations Memorial Cemetery in Korea(UNMCK)

UN Memorial Cemetery in Korea located in Busan is the only UN military cemetery in the world. The area is 13,3701 square meters, the size of 18 soccer fields combined.

재한 유엔기념공원

부산시에 자리한 재한 유엔기념공원은 세계에 단 하나밖에 없는 유엔군묘지입니다. 면적은 13만 3,701m㎡로 축구장 열여덟 개를 합친 크기입니다.

The UN Memorial Cemetery was established in January 1951 by the United Nations Command to bury the dead. After the war ended in November 1955, the National Assembly of the Republic of Korea donated the land here permanently to the United Nations. It was to repay the sacrifices of the UN forces. A month later, the United Nations General Assembly adopted a resolution to give the United Nations permanent management of the cemetery. Since November 1959, the United Nations Commission on the Reunification of Korea (UNCURK, Unkirk) has managed this facility.

유엔기념공원은 1951년 1월, 전사자 매장을 위하여 유엔군사령부가 조성하였습니다. 전쟁이 끝난 1955년 11월에 대한민국 국회는 이곳 토지를 유엔에 영구히 기증했습니다. 유엔군의 희생에 보답하기 위해서였어요. 한 달 후 유엔총회에서는 이 묘지를 유엔이 영구적으로 관리하기로 한 결의문을 채택했습니다. 1959년 11월부터 유엔 한국통일부흥위원단(UNCURK, 언커크)이 이 시설을 관리했고 (언커크가 해체된 1974년 이후부터는 11개국으로 구성된 유엔기념공원 국제관리위원회가 관리 업무를 맡고 있습니다).

During the Korean War, 40,896 UN soldiers were killed. Between 1951 and 1954, more than 11,000 remains were buried. Afterwards, all the remains of soldiers from seven countries, including Belgium, Colombia, Ethiopia, Greece, Luxembourg, the Philippines and Thailand, and some soldiers from other countries were transferred to their homeland.

6 · 25전쟁 기간 4만 896명의 유엔군이 희생되었습니다. 1951~1954년 사이에는 1만 1,000여 구의 유해가 여기 안장되어 있었어요. 그 후 벨기에, 콜롬비아, 에티오피아, 그리스, 룩셈부르크, 필리핀, 태국 등 7개국 장병들의 유해 전부와 그 외 국가의 일부 장병들의 유해가 그들의 조국으로 옮겨졌습니다.

Currently, the remains of 2,300 wards from 11 countries, including 36 Korean soldiers who died while dispatched to the UN forces, are enshrined. This number includes those who died before and after the war, such as husbands and wives. Each gravestone has a small bronze plate slightly larger than A4 paper engraved with nationality, name, affiliation, date of birth, and time of transcription.

현재는 유엔군 부대에 파견 중 전사한 한국군 36명을 포함하여 11개국 2,300여 구의 유해가 안치되어 있습니다. 이는 부부 합장자 등 전쟁 후에 사망 사람들을 포함한 숫자입니다. 각 묘소의 작은 비석에는 국적, 이름, 소속, 생년월일, 전사 시기 등을 새긴 A4 용지보다 조금 큰 동판이 붙어 있습니다.

The UN Memorial Cemetery has memorial towers and monuments for each

country that participated in the war. New Zealand erected a stone pagoda bearing symbol of the country, a fern, and Canada erected a statue of a soldier with two Korean children. In front of the statue is the inscription 'We do not the brave sons of Canada'. The children are holding twenty-one maple leaves and mugunghwa, representing the 21 missing Canadian soldiers. At the British military cemetery, there is a statue of a black lion commemorating the fallen soldiers and the rose, the national flower of England, is in full bloom.

유엔기념공원에는 각 참전국의 위령탑과 기념비가 있습니다. 뉴질랜드는 그 나라의 상징 식물인 고사리 문양을 담은 석탑을 세웠고 캐나다는 병사가 한국 아이 두 명을 데리고 있는 동상을 세웠어요. 동상 앞에는 '우리는 용감한 캐나다의 아들들을 잊지 않는다'라고 쓰여 있습니다. 아이들은 스물한 개의 단풍잎과 무궁화를 들고 있는데 이는 스물한 명의 캐나다군 실종자를 의미합니다. 영국군 묘지에는 전몰장병을 추모하는 흑 사자상이 있고 영국의 국화인 장미가 활짝 피어있습니다.

In the center of the UN Memorial Cemetery, there is a monument engraved with the names of all the fallen. The names are engraved on the black wall, and there are sculptures simulating the helmets of the dead, unquenchable flames, and walls of silence. At the UN Memorial Cemetery, there is a waterway called the Doeunt Waterway, and 'Doeunt' is the name of the youngest war veteran buried in the UN Memorial Cemetery. It was named in honor of a boy who came from Australia and died at the young age of 17. The Doeunt Waterway runs through the boundary between life (green area) and death (cemetery). The path of the Unknown Soldier continues in front of the Doeunt Channel.

유엔기념공원 중앙에는 모든 전사자의 이름을 새긴 추모 명비가 있습니다. 이름들은 검은 벽에 새겨져 있고 주변에 전사자의 철모를 형상화한 조각, 꺼지지 않는 불꽃, 침묵의 벽 등이 있습니다. 유엔기념공원에는 도트 수로라는 물길이 있는데 '도은트'는 유엔기념공원에 안장된 최연소 참전 용사의 이름입니다. 호주에서 와 17세의 어린 나이에 세상을 떠난 소년을 기리기 위해 붙인 이름이지요. 도은트 수로는 삶(녹지)과 죽음(묘역) 사이의 경계를 흐르고 있습니다. 도은트 수로 앞쪽으로는 무명용사의 길이 이어져 있습니다.

Every year on November 11th, at 11 am, the whole world pays a minute of silence towards Busan. This is a 'Turn toward Busan' event to commemorate the victims of the Korean War, which are buried in the UN Memorial Cemetery in Busan. It is a sure sign that people around the world still do not forget the Korean War and the veterans of the Korean War.

해마다 11월 11일 오전 11시가 되면 전 세계가 부산을 향해 1분간 묵념을 합니다. 이는 부산 유엔기념공원에 안장된 6 · 25전쟁 유엔군 전사자들을 위해 묵념하는 '턴 투워드 부산' 행사입니다. 전 세계 사람들이 아직도 6 · 25전쟁과 참전용사들을 잊지 않고 있다는 확실한 표시입니다.

United Nations Peace Memorial Hall

The UN Peace Memorial Hall in Busan was built for the purpose of showing to the generation who have not experienced war how miserable war is, how important the armistice is, and to thank the veterans for their sacrifice. The UN Peace Memorial Hall is carrying out various projects such as data collection and research, exhibitions, and commemorative events to enhance the honor of the UN Allied Nations and veterans. If you come up to the observatory of the UN Peace Memorial Hall you can see the UN Memorial Cemetery at a glance.

유엔평화기념관

부산에 있는 유엔평화기념관은 전쟁을 겪지 않은 세대에게 전쟁이 얼마나 참혹한 것인지, 정전협정이 얼마나 중요한 것인지를 알리고, 참전 용사들의 희생에 감사하는 목적으로 만들어 졌습니다. 유엔평화기념관에서는 유엔군 참전용사와 참전국의 명예를 높이기 위한 자료수집, 조사, 전시, 기념행사 등 다양한 사업을 추진하고 있습니다. 유엔평화기념관 전망대에 오르면 유엔기념공원이 한눈에 내려다 보입니다.

Yeoncheon UN Forces Crematorium Facility

The UN military crematorium facility in Misan-myeon, Yeoncheon-gun, Gyeonggi- do is presumed to be a facility built for cremation of UN soldiers who died during the Korean War. The centerpiece of the crematorium facility is a chimney and cremation pit, which is about 10 meters high, made of stone and cement. Only the lower wall of the two buildings that would have stood next to each other in an 'L' shape around it remains.

연천 유엔군 화장장 시설

경기도 연천군 미산면에 있는 유엔군 화장장 시설은 6 · 25전쟁 당시 전사한 유엔군을 화장하기 위해 만든 시설로 추정됩니다. 화장장 시설의 중심은 돌과 시멘트로 쌓은 10여m 높이의 굴뚝과 화장 구덩이입니다. 그 주변에 'L' 자 형태로 맞붙어 서 있었을 건물 두 동은 아래쪽 벽체만 남아 있어요.

It is still unclear whether this building was really built as a crematorium or was used as a crematorium. However, it is true that during the Korean War, a battle in hills were fiercely fought in the area around this facility, and many UN forces were killed. So the claim that this site is a crematorium facility is weighted

이 건물이 정말 화장장으로 지어졌는지, 화장장으로 쓰였는지 등에 대해서는 아직도 확실히 밝혀지지 않았습니다. 하지만 6 · 25전쟁 때 이 시설의 주변 지역에서 고지전이 치열하게 전개되어 유엔군이 많이 전사한 것은 사실입니다. 그래서 이 유적이 화장장 시설이라는 주장에 무게가 실리고 있습니다.

Reference

6 · 25전쟁에 두 번 참전, 전사한 그 청년[14]

7월27일(July 27) 6 · 25전쟁 정전협정 서명 70주년이 되는(mark the 70th anni-versary of the signing of the Korean armistice) 날이다. 미국 일간지 '산타페 뉴멕시칸'은 "그 전쟁을 치르러 부름받았던 이들 대부분은 이제 가고 없지만(most who were called to do the fighting are gone now), 그들의 이야기는 살아남아야 한다(their stories should live on)"며 뉴멕시코주(州) 데밍(Deming)에서 나고 자랐던 한 청년 얘기를 전했다.

"20세였다. 해병대 상병(Marine corporal) 클라이드 킬은 6 · 25 발발 몇 주 만인 8월 10일 한국에 도착했다. 그가 속한 중대는 10월 3일 4개의 동굴을 폐쇄하라는 임무를 수행하던 중(on a mission to seal four caves)교전을 벌이게 됐다(become engulfed in a firefight). 그는 반격에 주도적 역할을 했다(play a leading role in the counterattack). 적의 사격에 의도적으로 자신을 노출시켜(intentionally expose himself to enemy fire) 공격을 집중시킴으로써 아군 사상

14) 조선일보 A33(2023.07.25.)

자들을 후송하게 하고(allow casualties to be evacuated) 그사이에 동료들이 반격을 가할(mount a counteroffensive in the meantime) 수 있게 했다.

그는 이후에도 여러 무공을 세워 (distinguish himself in war) 1951년 3월 동성(銅星)무공훈장을 받았고, 그 무렵 복무 기간(tour of duty)이 끝나 제대하고(be dis-charged) 귀국하게 됐다. 그런데 얼마 후, 무슨 생각에서인지 그는 자원해서 해병대에 재입대해(volunteer to reenlist for some reason)다시 한국으로 보내달라고 요청했다. 그것이 생존자의 죄책감(survivor's guilt) 때문이었는지, 여전히 한국에서 목숨을 걸고(risk their lives) 싸우는 동료들과 함께하기를 원해서였는지, 정확히 아는 사람은 없다.

하사관으로 한국에 돌아온 그는 1953년 2월 3일 또 다른 격전에서 주도적 역할을 했다(play a leading part in another ferocious battle). 적의 강력한 박격포, 대포, 소총 공세를 무릎쓰고(in the face of intense enemy mortar, artillery and small arms fire) 산등성이를 타고 올라가 적의 참호와 벙커를 진압해 (clear out the enemy trenches and bunkers)목표 지점 꼭대기까지 점령했다. 또 작전상 후퇴를 할(make a strategic retreat) 때는 고지를 수도 없이 오르내리며 아군 사상자들을 안전한 장소까지 업어 날랐다.

지난 참전 때보다 더 높은 은성(銀星) 훈장을 받았다(receive the Silver Star). 6·25 전쟁에 두 번이나 참전한 사연과 함께 알려지면서 1953년 7월 17일 미국 신문에도 보도됐다. 그는 그 신문 보도가 나온 바로 그날 전투 중 전사했다(die in combat that very day). 23세 생일 하루 전날(the eve of his 23rd birthday)이었다. 6·25 정전협정 10일 전이었다."

Appendix

1. English Abbreviation Index List
2. 기수, 서수, 요일과 달, 시각
3. 사자성어 표현하기

Reference

1. English Abbreviation Index List

A

A/C	Aircraft	항공기
AA	Avenue of Approach	접근로
AA	Assembly Area	집결지
AAO	Air Assault Operation	공중강습작전
AAR	After Action Review	사후검토
ACA	Airspace Coordination Area	공역협조지역
ACM	Airspace Coordinating Measures	공역협조수단
ACP	Air Control Point	항공통제점
AD	Air Defense	방공
ADC	Area Damage Control	지역피해통제
ADC	Area Distribution Center	지역분배소
ADOA	Air Defense Operation Area	방공작전구역
AI	Air Interdiction	항공차단
AL	Airlift	공중수송
ALO	Air Liaison Officer	공군연락장교
AMC	At My Command	지명사
AO	Area of Operations	작전지역
AP	Advance Party	선발대
AROE	Armistice Rules of Engagement	정전 시 교전규칙
Arty	Artillery	포병
ARW	Air Raid Warning	공습경보
ASL	Authorized Storage List	인가저장품목
ASP	Ammunition Supply Point	탄약보급소
ATK	Attack	대화력전항공전력

ATO	Air Tasking Order	항공임무명령서
ATO	Anti-Terror Operation	대테러작전
ATP	Ammunition Transfer Point	탄약전환보급소
AUTL	Army Universal Task List	육군일반과업목록

B

B/L	Basic Load	기본휴대량
BCTP	Battle Command Training Program	전투지휘훈련
BDA	Battle Damage Assessment	전투피해평가
Bdry	Boundary	전투지경선
BHL	Battle Hand-over Line	전투이양선
BII	Basic Issue Item	기본불출품목
BMNT	Beginning Morning Nautical Twilight	해상박명초
BOQ	Bachelor Officer's Quarter	독신숙소
BSC	Battle Simulation Center	전투모의센터
BZ	Buffer Zone	완충지대

C

C	Confidential	Ⅲ급 비밀
C	Consumables	소모성품목
C/S	Chief of Staff	참모장
C4I	Command, COntrol, Communication, Computer & Intelligence	지휘, 통제, 통신, 컴퓨터 및 정보체계
CA	Civil Affairs	민사업무
CA	Combat Assault	강습
CA	Coordination Altitude	협조고도

CAO	Civil Affairs Operations	민사작전
CAS	Close Air Support	근접항공지원
CASOP	Crisis Action Standing Operating Procedures	위기조치예규
ccc	Command Control Center	지휘통제본부
CCIR	Command Critical Information Requirement	지휘관중요첩보요구
CCL	Civilian Control Line	민간인통제선
CCL	Combat Coordination Line	전투협조선
CD	Civil Defense	민방위
CDM	Chemical Downwind Message	화학풍통신문
CDR CFC	Commander, Combined Forces Command	연합군사령관
CEOI	Communications-Electronics Operation	통신전자운용지시
CEP	Circular Error Probable	원공산오차
CFFZ	Call for Fire Zone	사격요구지역
CFL	Coordinated Fire Line	사격협조선
CFX	Command Field Exercise	지휘조야외기동훈련
CFZ	Critical Friendly Zone	중요아군지역
CI	Counter Intelligence	대정보
CI	Command Inspection	지휘검열
CIPHER	Cipher Cryptogram	암호
Cmdr	Commander	지휘관
CMX	Crisis Management Exercise	위기관리연습
COA	Course of Action	방책
CODA	Combined Delegated Authority	연합권한위임사항
COP	Common Operations Picture	공통작전상황도
COP	Combat Outpost	전투전초
CP	Command Post	지휘소
CPMX	Command Post Movement Exercise	지휘소이동훈련
CPX	Command Post Exercise	지휘소훈련

C/S	Call Sign	호출호명
CSOA	Combined Special Operations Area	연합특수작전지역
CSR	Controlled Supply Rate	통제보급률
C T	Counter Terrorism	대테러

D

DCAO	Defensive Counter Air Operation	방어제공작전
DDR	Defensive Counter Air Operation Disarmament Demobilization	무장해제, 동원해제, 통합
DEFCON	Defense Condition	방어준비태세
DF	Direction Finding	방탐
D/L Equipment	Deadlined Equipment	불가동 장비
DM	Depot Maintenance	창정비
DMZ	Demilitarized Zone	비무장지대
DP	Decision Point	결심지점
DS	Direct Support	직접지원
DST	Decision Support Template	결심보조도표
DT	Developmental Test & Evaluation	개발시험평가

E

EA	Electronic Attack	전자공격
EBO	Effect Based Operations	효과중심작전
EDM	Effective Downwind Message	유효통신문
EEFI	Essential Elements of Friendly Information	작전보안핵심요소
EENT	End of Evening Nautical Twilight	해상박명초
ELINT	Electronic Intelligence	전자정보
EM	Emergency Mobilization	긴급동원
EMPB	Electro-Magnetic Pulse Bomb	전자기파탄
EOD	Explosive Ordnance Disposal	폭발물처리
EP	Electronic Protection	전자보호
EPW	Enemy Prisoner of War	포로
ER	Effective Range	유효사거리
ES	Electronic Warfare Support	전자전지원
EW	Electronic Warfare	전자전

F

FASCAM	Family of Scratterable Mines	살포식지뢰
FB	Forward Boundary	전방전투지경선
FD	Fire Direction	사격지휘
FEBA	Forward Edge of the Battle Area	전투지역전단
FSCL	Fire Support Coordination Line	화력운용협조선
FECM	Fire Employment Coordination Measure	화력운용협조수단
FEO	Forcible Entry Operations	강제진입작전
FFA	Free Fire Area	화력자유지역
FFIR	Friendly Force Information Requirement	우군정보요구
FLOT	Forward Line of Own Troops	전선부대진출선
FM	Field Manual	야전교범
FOUO	For Official Use Only	대외비
FR	Force Restoration	전투력복원
FT	Firing Tables	사표
FTX	Field Training Exercise	야외기동훈련

G

GOP	General Outpost	일반전초
GS	General Support	일반지원
GW	Guerrilla Warfare	유격전

H

HA	Humanitarian Assistance	인도적지원
HDM	Homeland Defense Mobilization	향방동원
HEAT	High Explosive Anti-Tank	대전차고폭탄
HOB	Height of Burst	파열고
HRP	Highway Regulation Point	육로조정소
HVT	High Value of Target	고가치표적

I

I/C	Interchangeable Capability	호환성품목
ICBM	Intercontinental Ballistic Missle	대륙간 탄도 미사일
ICL	Intelligence Coordination Line	정보협조선
ICV	Infatry Combat Vehicle	보병 전투차량
IDC	Integrated Defense Center	통합방위본부
IDS	Integrated Defense Situation	통합방위사태
IED	Improvised Explosive Device	급조폭발물
IFF	Identification Friend or Foe	피아식별
ILS	Integrated Logistics Support	종합군수지원
IN	Infantry	보병
IO	Information Operations	정보작전
IP	Intelligence Production	정보생산
IPB	Intelligence Preparation of the Battlefield	전장정보분석
IR	Intelligence Requirements	정보요구
IR	Information Requirements	첩보요구
ISR	Intelligence Surveillance Reconnaissance	정보감시정찰

J

JAAT	Joint Air Attack Team	합동공중공격반
JAGC	Judge Advocate General's Corps	법무
JAOP	Joint Aerial Observation Post	합동대공감시소
JFA-K	Joint Fires Area-Korea	합동화력지역
JMS	Joint Military Strategy	합동군사전략서
JPITL	Joint Prioritized Integrated Target List	합동우선순위통합 표적목록
JSCP	Joint Strategic Capabilities Plan	합동군사전략 능력기획서
JSOP	Joint Strategic Objective Plan	합동군사전략 목표기획서

K

KADIZ	Korea Air Defence Identification Zone	한국방공식별구역
KAOC	Korea Air and Space Operations Center	한국항공우주작전본부
KLIZ	Korea Limited Identification Zone	한국제한식별구역

L

LC	Line of Contact	접촉선
LD	Line of Departure	공격개시선
LOC	Line of Communication	병참선
LOS	Line of Sight	조준선
LOTS	Logistics Over-The-Shore	해양양륙군수지원
LZ	Landing Zone	착륙지대

M

MASINT	Measurement & Signature Intelligence	계측 및 기호정보
M-DAY	Mobilization Day	동원선포일
MDL	Military Demarcation Line	군사분계선
MEMO	Mission Essential Maintenance Only	임무필수정비
METL	Mission Essential Task List	임무필수과업목록
METT+TC	Mission, Enemy, Terrain and Weather Troops Available, Time Available, Civil Considerations	전술적고려요소, 임무, 적, 지형 및 기상, 가용부대, 가용시간, 민간요소
MISO	Military Information Support Operations	군사정보지원작전
MLC	Martial Law Command	계엄사령부
MLE	Minimum Line of Expect	최소예상선
MLSA	Mutual Logistics Support Agreement	상호군수지원협정
MM	Mobile Maintenance	이동정비
MOPP	Mission Oriented Protective Posture	임무형보호태세
MP	Military Police	군사경찰
MRL	Minimum Resistance Line	최소저항선
MRR	Minimum Risk Route	최소위험경로
MSEL	Master Scenario Event List	사태목록
MSR	Main Supply Route	주보급로

N

NAI	Named Area of Interest	중요감시지역
NBC	Nuclear Biological and Chemical	화생방
NCU	Normal Call Up	정상동원
NEO	Noncombatant Evacuation Operations	비전투원후송작전
NFA	No Fire Area	화력금지구역
NFL	No Fly Line	비행금지선
NLL	Northern Limit Line	북방한계선

O

O	Organization	편성
O B	Order of Battle	전투서열
OCAO	Offensive Counter Air Operations	공세제공작전
OEM	ON Equipment Material	장비부수기재
OP	Orbit Point	선회대기지점
OPCOM	Operational Command	작전지휘
OPCON	Operational Control	작전통제
OPLAN	Operation Plan	작전계획
OPORD	Operation Order	작전명령
OPSEC	Operations Security	작전보안

P

PE	Probable Error	공산오차
PIR	Priority Intelligence Requirements	우선정보요구
PKO	Peace Keeping Operation	평화유지작전
PL	Passage of line	초월선
PL	Phase Line	통제선
PM	Preventive Maintenance	예방정비
Pre-ATO	Prepositional Air Tasking Order	기계획항공임무명령서

Q

QM	Quartermaster	병참
QRF	Quick Reaction Forces	기동타격대

R

R&D	Research and Development	연구 및 개발
R&S	Reconnaissance and Surveillance	정찰감시
RC	Reserve Components	예비군
RCN	Reconnaissance	정찰
REP	Range Error Probable	사거리공산오차
RF	Reinforcing	증원, 화력증원
RFA	Restriction Fire Area	화력제한지역
RFI	Request for Information	첩보요청
RI	Relevant Information	관련정보
RM	Recurring Maintenance	순환정비
ROC	Requirement Operation Capability	작전운용성능
ROTC	Reserve Officers Training Corps	학생군사훈련단(학군단)

ROP	Rear Orbit Point	후방대기지점
ROZ	Restricted Operation Zone	작전제한지역
RP	Reporting Post	전방감시소
RSOI	Reception Staging, onward Movement and Integration	연합전시증원 수용, 대기, 전방이동 및 통합
RSR	Required Supply Rate	소요보급률
RT-Box	Artillery Box	포병박스

S

SC	Strategic Communication	전략커뮤니케이션
SEAD	Suppression of Enemy Air Defence	적 방공제압
SIGINT	Signals Intelligence	신호정보
SIR	Specific Information Requirement	세부첩보요구
SITREP	Situation Report	상황보고
SLBM	Submarine Launched Ballistic Missile	잠수함 발사 탄도미사일
SR	Situation Report	상황보고
SOP	Standing Operating Procedure	예규, 내규
SSS	Special Air Surveillance Sector	특별공중감시구역

T

T&E	Test and Evaluation	시험평가
T/D & A	Table of Distribution & Allowance	분배 및 배당표
T/O	Table of Organization	편제표
T/O & E	Table of Organization & Equipment	편성 및 장비표
TACON	Tactical Control	전술통제
Tag	Target	표적
TAI	Target Area of Interest	관심타격지역
TAL	Tactical Action line	전술조치선
TC	Training Circular	교육회장
TF	Task Force	특수임무부대
TOT	Time on Target	동시탄착사격
TPFDD	Time-Phased Force and Deployment Data	시차별부대전개제원

U

UAV	Unmanned Aerial Vehicle	무인정찰기
UFS	Ulchi-Freedom Shield	을지자유의방패 연습
USFK	US Forces, Korea	주한미군
UTM Grid	Universal Transverse Mercator Grid	유티엠 좌표

V

VTC	Video Tele-Conference	화상회의

W

WATCHCON	Watch Condition	정보감시태세
WCS	Weapons Control Status	무기통제상태
WFF	War Fighting Function	전투수행기능
WHNS	Wartime Host Nation Support	전시주둔국지원
WMD	Weapons of Mass Destruction	대량살상무기
WRSA	War Reserve Stocks for Allies	동맹국전쟁예비비축물자
WS	Weapon Systems	무기체계

X

X-ATK	Airborne-Alert Attack	공중비상대기화력항공전력
X-INT	Airborne-Alert Interdiction	공중비상대기대화력전항공차단

Y

YDT	Yard Diving Tender	구조지원정

2. 기수, 서수, 요일과 달, 시각, 숫자

1. 기수와 서수

- 기수는 사물의 개수나 양을 나타내는 수.
- 서수는 사물의 순서를 나타내는 수.
- 수 '1, 2, 3, 4, 5…'는 수가 나타내는 의미에 따라 읽는 방법이 다르다. 개수를 나타낼 때에는"하나, 둘, 셋…"으로 읽어야 하고, 차례를 나타낼 때에는"첫째, 둘째, 셋째…" 또는 "일, 이, 삼…"으로 읽는다.

구분	기수	서수	구분	기수	서수
1	one	first	21	twenty-one	twenty-first
2	two	second	22	twenty-two	twenty-second
3	three	third	23	twenty-three	twenty-third
4	four	fourth	24	twenty-four	twenty-fourth
5	five	fifth	25	twenty-five	twenty-fifth
6	six	sixth	26	twenty-six	twenty-sixth
7	seven	seventh	27	twenty-seven	twenty-seventh
8	eight	eighth	28	twenty-eight	twenty-eighth
9	nine	ninth	29	twenty-nine	twenty-ninth
10	ten	tenth	30	thirty	thirtieth
11	eleven	eleventh	40	forty	fortieth
12	twelve	twelfth	50	fifty	fiftieth
13	thirteen	thirteenth	60	sixty	sixtieth
14	fourteen	fourteenth	70	seventy	seventieth
15	fifteen	fifteenth	80	eighty	eightieth
16	sixteen	sixteenth	90	ninety	ninetieth
17	seventeen	seventeenth	100	hundred	hundredth
18	eighteen	eighteenth	1000	thousand	thousandth
19	nineteen	nineteenth	-	-	-
20	twenty	twentieth	-	-	-

2. 요일과 달

요일	월 Monday	화 Tuesday	수 Wednesday	목 Thursday	금 Friday	토 Saturday	일 Sunday

달						
	1월 January	2월 February	3월 March	4월 April	5월 May	6월 June
	7월 July	8월 August	9월 September	10월 October	11월 November	12월 December

3. 시각 묻고 답하기

(1) 몇 시입니까?
- What time is it?(What time ≠ When)

= What's the time?

= What time do you have?

= Do you have the time? = Are you free?

(2) 몇 시 몇 분입니다.
- It is 0시 0분.

= Are you free?

= It is 0분 전 0시.

= It is 0분 지난 0시.(전: before, to / 지난: after, past)

[참고] • a quarter: 15분 • half: 30분

- It's two five. = It's five after two.
- It's ten fifty. = It's ten before eleven.
- It is seven thirty. = It is half past seven.
- It is eight fifteen. = It is a quarter to eight.

4. 숫자 읽는 방법

- 큰 수: 세 자리씩 끊어 읽는다.
  - 10,000 - ten thousand
  - 1,000,000 - one million
- 연도 : 두 자리씩 나누어 읽는다.
  - 1800 - eighteen hundred
  - 1995 - nineteen ninety-five
- 월, 일 : 보통 서수로 읽는다.
  - 3월 10일 - March (the) tenth = March ten = the tenth of March
- 전화 번호: 한 자리씩 읽는다.
  - 302-1234 - three o(zero) two, one two three four
- 주요고지 및 도로망: 한 자리씩 읽는다.
  - 405고지 mountain four o(zero) five
  - 202번 도로 road number two o(zero) two
- 화폐
  - $4,30 - four dollars (and) thirty (cents)

3. 사자성어 표현하기

가화만사성

집안이 화목하면 모든 일이 잘 이루어진다는 것
when one's home is harmonious, all goes well.
he is the happiest, be he king or peasant, who finds peace in his home.

갑론을박

to argue the pros and cos.
you argue for and I will argue against.

감언이설

남의 비위를 맞추는 달콤한 말과 이로운 조건을 들어 그럴듯하게 꾸미는 말.
flattery will get you nowhere.
flattery will get you everywhere.

견물생심

물건을 보면 그것을 가지고 싶은 욕심이 생김.
seeing is wanting.

결자해지

매듭은 맺은 사람이 풀어야 한다는 뜻.
you have to live with the consequences.
you have made your bed, now lie on it.

구사일생

that was a chose shave.

that was a narrow escape.

마이동풍

아무리 얘기해도 소용이 없다.

to fall on deaf ears.

to go unheeded.

my advice fell on deaf ears.

명약관화

불을 보듯이 명백함.

crystal clear.

부화뇌동

우레 소리에 맞춰 함께 한다는 뜻으로, 자신의 뚜렷한 소신 없이 남이 하는 대로 따라감.

follow blindly.

follow suit.

copy cat.

작심삼일

a short-lived resolution.

삼고초려

인재를 맞기 위해 여러 번 찾아가서 예를 다하는 일.

ex) they visited Mr.Lee three times and the third time was the charm.
third time's a charm; third time's the charm.

신상필벌

상을 줄 사람에게 상주고 벌 줄 사람 벌주는 것
just reward.
reward a good deed and punish a misdeed.

심기일전

어떤 동기에 의하여 지금까지 품었던 생각과 마음의 자세를 완전히 바꿈.
to do an about face.
I decieded to turn over a new leaf.

안하무인

눈에 보이는 사람이 없음
self-centered.
arrogant, disrespectful.

언중유골

말에 뼈가 있다.
말 속에 숨겨진 의미가 있다.
the implied meaning in a remark.

연목구어

나무에 올라가 고기를 얻으려고 한다.
you are baking up the wrong tree.
you are knocking on the wrong door.

이심전심

you took the words right out of my mouth.
we are on the same wavelength.

이전투구

볼썽사납게 서로 헐뜯거나 다투는 모양을 비유함.
to fight like cats and dogs.

일파만파

to have a ripple effect.

오리무중

I have a no idea.
I'm in a fog.

오월동주

서로 적의를 품은 사람들이 같은 처지나 한 자리에 있게 됨을 비유함.
bitter enemies placed by fate in the same boat.

오비이락

새가 날자 배가 떨어진다, 우연의 일치
a pure coincidence.

우왕좌왕

he was running around like a chicken that's lost its head.

유비무환

방비가 잘되어 있으면 적이 언제 쳐들어와도 근심이 없음.
forewarned is forearmed.
an ounce of prevention is worth a pound of cure.

자업자득

자기가 저지른 일의 과오를 자기자신이 받음
to reap the fruits of your actions.
you pay your money and you take your chances.

적반하장

the thief turns on the master with a club.

재승박덕

재주는 있으나 덕이 적음
to be talented and yet possess little virtue.

주마간산

대충 보다
to skim through, to flip through.
to glance at.

천지개벽

자연계나 사회의 큰 변동
a cataclysmic change.

초지일관

accomplish one's original goals.
fulfill one's original intentions.

파죽지세

대나무를 쪼갤 듯한 기세
the troops advanced unopposed.
the floods swept away everything in its way.

풍전등화

위급한 처지를 말함
his fate hangs in the balance.
his fate hangs by a thread.

화룡점정

가장 요긴한 부분을 마치어 일을 끝냄.
the finishing stroke / touch

Reference

김성우 · 정기주, 『Low & High Level Military English』(인천: 진영사, 2016.03.01.)

국방부, 『국방백서 2014』(국방부, 2014.12.31.).

국방부, 『국방백서 2018』(국방부, 2018.12.31.).

국방부, 『국방백서 2018』(영문판, 국방부, 2018.).

국방어학원, 『M-MOOC 군사영어 보충교재』(국방어학원, 2021.03.02.).

박휘락, 『국가안보론』(경기: 도서출판 대영문화사, 2020).

육군사관학교 영어과, 『Military English Ⅰ』(서울: 도서출판 봉명, 2008).

육군본부, 『군사용어』(육군본부, 2019).

육군본부 군사연구소, 『1129일간의 전쟁 6 · 25』(육군본부, 2014).

육군정보학교, 『Military English』(육군정보학교, 1999.07.10.).

연합사, 『Practical Military English Ⅱ』(비매품).

조용만 · 최병욱, 『Military English』 (경기: 북코리아, 2020.03.15.).

채서영, 『영어는 대체 왜? 그런가요』(서울: 사회평론, 2021.04.16.).

황인희, 『감사합니다, 잊지 않겠습니다 한권으로 읽는 유엔 참전국 이야기』(서울: 양문, 2023.04.11.).

FM 6-0, Commander and Staff Organization and Operations(Headquarters, Department of the Army, May 2014).

TRADOC Pamphlet 525-3-1, The US Army Operating Concept(October 2014).

집필자 약력

전규호 (국방안보학 박사)

청주대학교 법학과를 졸업하고 국방대학원 안보정책학 석사,
청주대학교에서 국방안보학 박사학위를 취득하였음.

육군 정보학교 군사영어반, 합참 해외파병과 작전담당관,
23사단 58연대장, 해군작전사 합동작전조정관, 육군본부 지휘통제실장 · 훈련과장,
한-튀르키예 육군회의 교육훈련분과 위원장, 청주대학교 학군단장,
강동대학교 군무원 부사관학과 교수를 역임하였음.

현재 청주대 군사학과 교수로 재직 중에 있으며,
군사영어, 국가안보론, 대테러작전, 리더십을 주로 강의함.

정기주 (행정학 박사)

육군 정보학교 군사영어반 수료
군사령부 인사 · 작전참모부 실무자
한 · 미 연합작전 통역담당

부경대학교 학생군사교육단 교관
부경대학교 대외사업팀장

현재 동명대학교 학생군사교육단 교수,
유엔평화기념관 연구위원